BATCH PROCESSING

Modeling and Design

BATCH PROCESSING

Modeling and Design

URMILA DIWEKAR

CRC Press
Taylor & Francis Group
Boca Raton London New York

CRC Press is an imprint of the
Taylor & Francis Group, an **informa** business

CRC Press
Taylor & Francis Group
6000 Broken Sound Parkway NW, Suite 300
Boca Raton, FL 33487-2742

First issued in paperback 2017

Version Date: 20131227

ISBN 13: 978-1-138-07674-7 (pbk)
ISBN 13: 978-1-4398-6119-6 (hbk)

Visit the Taylor & Francis Web site at
http://www.taylorandfrancis.com

and the CRC Press Web site at
http://www.crcpress.com

To my sister Utpala Diwekar, who taught me how to study.

Contents

List of Figures

List of Tables

Preface

Batch processes are widely used in pharmaceutical, food, and specialty chemicals where high value, low volume products are manufactured. Recent developments in bio-based manufacturing also favor batch processes because feed variations can be easily handled in batch processes. Further, the emerging area of nanomaterials manufacturing currently uses batch processes as they are low volume, high energy intensive processes. Design of these processes is difficult because of the time dependent nature of the process and the allowable flexibility. Although batch processing has existed for a long time, designing these processes and unit operations was an onerous task and required computational efforts. Therefore, traditional design books did not cover batch processing in detail. The aim of this book is to fill this void. The book describes various unit operations in batch and bio-processing and design methods for these units. Numerical methods are necessary to solve these design problems. The time dependent nature of batch processes result in challenging scheduling and planning problems. Further, every unit encounters optimal control problems. This book provides basics to solve these problems. The book is intended as a text as well as reference book for researchers and practitioners alike.

Contributors

- Dr. Urmila M. Diwekar, Vishwamitra Research Institute, Clarendon Hills, IL 60514

- Dr. Saadet Ulas Acikgoz, UOP, A Honeywell Company, Des Plaines, IL 60017

- Dr. Demetri Petrides, Douglas Carmichael, and Charles Siletti, Intelligen Inc., Scotch Plains, NJ 07076

Acknowledgments

I want to dedicate this book to my sister Utpala Diwekar. Right from my first school day, she was encouraging me to study and acquire knowledge. Thanks also to my youngest niece Ananya Joshi, who asked me to dedicate this book to her favorite aunt, Raju-mawshi.

I am very appreciative of the efforts of Dr. Vicente Rico-Ramirez, my collaborator from Mexico, my post-doctoral fellow Dr. Berhane Gebreslassie, and my graduate students Pahola Benavides and Kirti Yenkie for the careful review of the initial drafts of the manuscript and for their comments.

Of course, I am especially thankful to my dearest and nearest family members Anjali Diwekar, Ashish Joshi, and Sanjay Joag for their support and patience. Their enthusiasm and admiration of even very small advancements in my research and other endeavors made all the efforts worthwhile.

To my parents, special thanks for their untiring love and caring, and for all the sacrifices behind their greatest gift to me—education.

Urmila M. Diwekar
Clarendon Hills, IL
August, 2013

1

Introduction

CONTENTS

1.1 Introduction

Batch processes are widely used in pharmaceutical, agricultural, commodity food, and biotechnology bio products industries. These industries account for several billion dollars in annual sales. The cost of developing bio and pharmaceutical products is considerably high. However, there are a relatively small number of books which are dedicated to designing and operation of these processes. Thus, the following chapters provide process design principles for unit operations involved in these processes. Typically these unit operations involve time dependent processes and have to deal with differential equations which need to be integrated numerically for most of the cases. Therefore, Chapter 2 describes typical numerical integration techniques that can be used to solve the design and operation differential equations of these unit operations. The techniques covered include integration explicit methods starting with Euler's method to implicit methods for stiff systems like the backward difference formula. The following chapters use these methods for simulation of various batch unit operations. Material and energy balances, kinetics, equilibrium, and transport equations form the scientific foundation for engineering calculations. Reactions are the heart of any production, therefore, Chapter 3 is devoted to batch and semi-batch reactors like fed-batch reactor. Various types of reactions are discussed with the emphasis on autocatalytic, and parallel and series reactions which are common in bio processing.

1.2 Batch Separation

Separation unit operations form the rest of the process in batch processing. Separation processes depend on property differences and can be categorized into two categories, namely, contact equilibrium separation processes and mechanical separation processes. Distillation, absorption, extraction, adsorption and chromatographic separation, crystallization, and drying are based on thermodynamic equilibrium and can be categorized as contact equilibrium processes. Whereas filtration and centrifugation are mechanical separation processes. The chapters on separation processes are described in the following order:

- Batch Distillation: Chapter 4 is devoted to batch distillation. This is one of the most important and one of the most studied unit operations in batch industries. Separation is based on vapor-liquid equilibria. There are a number of configurations possible in conventional batch column, namely, the constant reflux mode, the variable reflux mode, and the optimal reflux mode. There are a number of new configurations that have emerged in the literature for batch distillation. This chapter describes all these operating modes and configurations. Various levels of models are available for different analysis. Different numerical integration techniques are needed to solve equations of these different models. Optimization and optimal control are well studied for this unit operation.

- Batch Absorption: Chapter 6 describes briefly the batch absorption operation. This unit operation is based on the solubility difference. Batch absorption is mostly used for finding kinetics and mass transfer coefficients for the absorption process.

- Batch Extraction: Chapter 7 presents the batch extraction operation. There are two types of extraction. Solid extraction or leaching involves solids which are leached or extracted by solvent where solubility of solids is the important property used for separation, and liquid extraction where immiscible solvent is used to extract liquid product from the mixture using liquid-liquid equilibria.

- Batch Adsorption: Chapter 8 describes batch adsorption which is based on differential affinity of various soluble molecules to solid adsorbents. The equilibrium relationship between solid, liquid, or gas phase is determined using adsorption isotherms.

- Batch Chromatography: Chromatographic separation is based on the same principle as adsorption. Most commonly this unit operation is used to measure a wide variety of thermodynamic, kinetic, and physico-chemical properties.

- Batch Crystallization: Chapter 10 is devoted to batch crystallization where a phase diagram is used to find the supersaturation at which point material crystallizes. This is again one of the most studied batch operations. Similar to batch distillation, various modeling techniques are used to describe the operation of batch crystallizer, and optimization and optimal control problems are well studied.

- Batch Drying: Chapter 11 describes the drying operation. Drying mostly involves removal of water and hence the phase diagram of water is important to describe this operation.

- Batch Filtration: Filtration involves separation of solids from liquid using a porous medium or screen. A mechanical separation process, this unit operation involves less energy than drying.

- Batch Centrifugation: Chapter 13 is devoted to batch centrifugation where centrifugal force is used for separation. Centrifugation is used for separating particles on the basis of their size or density, separating immiscible liquids of different densities, filtration of a suspension, breaking down of emulsions and colloidal suspensions, or for separation of gases, e.g., isotope separation in nuclear industry.

1.3 Optimization and Optimal Control

The time dependent nature of unit operations in batch processing pose challenging optimization and optimal control problems for units. In general, control refers to a closed-loop system where the desired operating point is compared to an actual operating point and a knowledge of the error is fed back to the system to drive the actual operating point towards the desired one. However, the optimal control problems we consider here do not fall under this definition of control. Because the decision variables that will result in optimal performance are time-dependent, the control problems described here are referred to as optimal control problems. Thus, use of the control function here provides an open-loop control. The dynamic nature of these decision variables makes these problems much more difficult to solve as compared to normal optimization where the decision variables are scalers. In batch processing scheduling and planning encounter various optimization problems where the decision variables are scalers. Therefore, a separate chapter (Chapter 5) on basics of optimization and optimal control problem formulation and solution method is presented. This rationale behind including this chapter in the middle of the book instead of in the begining is to give readers some familiarity with the optimal control problems involved in basic unit operations like batch reactor and batch distillation before introducing the methods to solve it.

The last but one chapter in this book is devoted to batch scheduling and planning. These problems are part of optimization problems and hence an industrial case study illustrate how to formulate such problems and solve it. Batch process simulation is the last chapter, mostly devoted to batch process simulation software and illustrative case studies.

1.4   Summary

This chapter provided the introduction to various unit operations described in this book. Introduction to solution techniques involved in solving differential equations, optimization, and optimal control problems commonly encountered in batch processing is also presented.

2

Numerical Methods for Integration

CONTENTS

Batch and bio process modeling often results in a set of ordinary differential equations. Numerical solution techniques for solving these differential equations normally involve approximating the differential equations by difference equations that are solved in a step-by-step marching fashion.

For example, consider the following system of equations.

$$\frac{d\bar{y}(\bar{x})}{d\bar{x}} \;=\; f(\bar{x}, \bar{y}(\bar{x})) \qquad (2.1)$$

Here $\bar{y}$ and $\bar{x}$ denote the vector of variables. Any numerical integration scheme for the equation shown above can be represented in the following general form where h denotes the step size and $k + 1$ represents the next step.

$$y_{k+1} \;=\; \sum_{i=1}^{m} \beta_i y_{k+1-i} \;+\; h\phi(x_{k+1}, \ldots, x_{k+1-m}, y_{k+1}, \ldots, y_{k+1-m}) \qquad (2.2)$$

where m is the order of the polynomial, ϕ is some functional.

In this chapter, we will present a number of methods for numerical integration of ordinary differential equations. We also present orthogonal collocation method for partial differential equations as well as ordinary differential equations. A major portion of this chapter is derived from the work by the author on batch distillation [7].

2.1   Error and Stability

A major question regarding any numerical integration technique is the accuracy of the approximation, which depends upon the errors, convergence, and stability of the integration techniques. In general, the error in applying these techniques comes from two sources: the truncation error that results from the replacement of the differential equation by the difference approximation (η), and the rounding error (ϵ) made in carrying out the arithmetic operations of the method. The above approximation can be rewritten using the error terms as follows.

Local truncation error:

$$y(x_{k+1}) - y_{k+1} = h\eta_k \tag{2.3}$$

rounding errors:

$$y_{k+1} = \sum_{i=1}^{m} \beta_i y_{k+1-i} + h\phi(x_{k+1}, \ldots, x_{k+1-m}, y_{k+1}, \ldots, y_{k+1-m}, \eta_k^1) + \epsilon_k^2 \tag{2.4}$$

Normally, the truncation error goes to zero as h approaches zero. Hence, we can make the error as small as we want by choosing an appropriate h. However, the smaller the h, the larger the rounding error on the computed solution. In practice for a fixed word length in the computer arithmetic, there will be a size h, below which the rounding error will become the dominant contribution to the overall error. Because a faster solution is always preferred, it is desirable to choose as big a step size as possible where the truncation error is predominant.

For a method to be convergent, the finite difference solution must approach the true solution as the interval or the step size approaches zero.

The concept of stability is associated with the propagation of errors of the numerical integration technique as the calculation progresses with a finite interval size, and is related to the effects of errors made in a single step on succeeding steps. The problem of stability arises because in most instances the order m of the approximate difference equation (Equation 2.4) is higher than the original differential equation q (Equation 2.1), whose actual solution can be written as $(m > q)$:

$$y(x) = \sum_{i=1}^{q} \gamma_i g_i(\bar{x}) \tag{2.5}$$

Hence, the difference equation may contain extraneous terms that may dominate the equation and bear little or no resemblance to the true solution. It happens frequently that the spurious solutions do not vanish even in the limits as the increment size approaches zero. This phenomenon is called *strong instability*, and implies lack of convergence as well as lack of stability. When a method possesses convergence but has unstable asymptotic behavior, the phenomenon is called *weak instability*. For example, a numerical routine that is stable for some finite increment size, such as h_1, but is unstable for some larger increment size, e.g., h_2, $(h_2 > h_1)$, is said to possess a weak instability ([8]). Even if the order of the approximate difference equation is less than or equal to the order of the original differential equation (e.g., lower order methods, such as Euler's integration, which is discussed in the next subsection), the method can be stable over a very small interval. On a finite interval, a stable method gives accurate results for a sufficiently small h. On the other hand, even strongly stable methods can exhibit unstable behavior if h is too large. Although in principle h can be taken to a sufficiently small value to overcome this difficulty, it may be that the computing time then becomes prohibitive. Furthermore, the rounding error can become dominant, resulting in an erroneous solution. In the next section, we will discuss different numerical integration techniques, their error and stability characteristics, and their applicability to the solution of batch process dynamics.

2.2   Numerical Integration Techniques

Numerical integration techniques can be categorized into two types:

- one-step methods

- multistep methods

Another way of classifying the integration techniques depends on whether or not the method is explicit, semi-implicit, or implicit. The implicit and semi-implicit methods play an important role in the numerical solution of stiff differential equations. To maintain the continuity of the section, we will first describe the explicit integration techniques in the context of one-step and multistep methods. The concept of stiffness and implicit methods are considered in a separate subsection, which also marks the end of this section.

2.2.1   One-Step Methods

In single-step methods at each step of integration the solution depends on the prior step only. The generalized form of a single-step method, as derived from equation 2.2, can be written as:

$$y_{k+1} = y_k + h\phi(x_k, y_k) \tag{2.6}$$

The functional form of ϕ changes according to different methods. The most commonly used methods in this category are Euler's method and Runge-Kutta methods. These methods are described below.

Euler's Method

Euler's method perhaps represents the simplest numerical integration technique, which can be defined as:

$$y_{k+1} = y_k + hf(x_k, y_k) \tag{2.7}$$

Euler's method is derived from a first order Taylor series approximation, where the higher derivative terms starting from the second derivative are ignored. It can be easily proved that the local truncation error, $L(h)$, for Euler's method decreases proportionally to the step size, i.e., $L(h) = O(h)$.

Runge-Kutta Methods

In Runge-Kutta methods, Euler's simple formula for calculating the functional derivative at one point and then using it to calculate the Taylor series expansion shown in equation 2.7, is replaced by derivative information at a number of points. As the number of points increases, the order of the method increases. The order of the method is defined to be the integer p for which $L(h) = O(h^p)$. In other words, for the $p-th$ order method, we expect the local truncation error to decrease by a factor of about 2^{-p} when we halve h, assuming that h is sufficiently small. So, for the fourth-order Runge-Kutta method in the following equation, the local truncation error has the form given by $L(h) = O(h^4)$.

Fourth-Order Runge-Kutta Method:

$$k_0 = hf(x_k, y_k)$$

$$k_1 = hf(x_k, y_k + \frac{k_0}{2})$$

$$k_2 = hf(x_k, y_k + \frac{k_1}{2})$$

$$k_3 = hf(x_k, y_k + k_2)$$

$$y_{k+1} = y_k + \frac{1}{6}(k_0 + 2k_1 + 2k_2 + k_3) \tag{2.8}$$

Higher order Runge-Kutta methods have a better stability range than Euler's integration technique.

The one-step methods described above are the Taylor series approximation where the higher derivative terms are ignored. However, many times this approximation may not hold true, and variable stepsize methods are then preferable. In spaces where higher-order derivatives cannot be ignored for that step, use very small h, otherwise, use a larger step size. Several different variants of the methods presented here are available where higher order derivative information can be used to obtain automatic step size changes. For more details, refer to [9]-[10] .

2.2.2 Multistep Methods

In one-step methods, the integration formula depends on previous steps (i.e., $k+1$-th point is calculated from k-th point). However, since all these methods are dependent on polynomial approximations, the information about other points such as $k-1$, $k-2$, etc., is also included in the approximation which makes these methods more realistic. The multistep methods are dependent on more steps than just the previous one. The multistep methods are attractive because they provide better representation of the functional space and, hence, better accuracy. The following equation provides the generalized representation of the multistep methods where the function f is replaced by a polynomial function p.

$$y_{k+1} = y_k + \int_{x_k}^{x_{k+1}} f(x, y(x))dx = y_k + \int_{x_{k-j}}^{x_{k+1}} p(x)dx \qquad (2.9)$$

However, multistep methods suffer from two problems not encountered in one-step methods. One problem is associated with the starting of these methods. The $k+1$-th step is dependent on k and $k-1$ steps, etc., and at $k=1$ there is no information about these previous steps. To circumvent this problem, a one-step method is used as a starter for a multistep method, until all the information is gathered. Alternatively, one may use a one-step method at the first step, a two-step method at the second, and so on, until starting values have been built up. However, it is important that the starting method used in subsequent stages maintain the same accuracy in the initial stages as the multistep method (which means that, initially one must use a smaller step size).

The second problem with multistep methods is the presence of extraneous solutions. This comes under the category of techniques where the order of the solution method is more than the order of differential equations ($m > q$). Therefore while these methods provide greater accuracy, they may also possess strong instability characteristics. The following paragraphs describe some of the multistep methods.

Adams-Bashforth Methods

Euler's method can be considered as the simplest form of the Adams-Bashforth method, where the function f is represented by a polynomial p of order 1. If p is assumed to be a linear function that interpolates between (x_{k-1}, f_{k-1}) and (x_k, f_k), then p can be represented by:

$$p(x) = p_1(x) = f_k - \frac{x - x_k}{h}\delta f_k \qquad (2.10)$$

where $\delta f_k = f_{k-1} - f_k$ and $h = |x_{k-1} - x_k|$. Integrating the above equation and substituting it in equation 2.9 results in the two-step Adams-Bashforth method formula given below.

$$y_{k+1} = y_k + hf_k - \frac{h}{2}\delta f_k = y_k + \frac{h}{2}(3f_k - f_{k-1}) \tag{2.11}$$

Similarly, one can obtain higher-order Adams-Bashforth methods by using higher-order polynomials.

The truncation error for a $p - th$ order multistep method is the same as that obtained by a $p-th$ order one-step method, i.e., $L(h) = O(h^p)$. However, in multistep methods higher-order methods can be constructed by evaluating f once per step.

The Adams-Bashforth methods use information about prior points. In principle, one can form polynomials using forward points as well. Using the points x_{k+1}, x_k, $\ldots$, x_{kN} to form a $N + 1$ polynomial generates a class of methods known as Adams-Moulton Methods. However, in these methods also calculation of y_{k+1} requires the solution of f_{k+1} implicitly. Implicit methods are discussed separately in a section which deals with stiff equations. One can also use a combination of an implicit method, such as an Adams-Moulton method, along with an explicit method, like an Adams-Bashforth method, to form an explicit method known as the Predictor-Corrector Method, which is discussed below.

Predictor-Corrector Method

The most commonly used Predictor-Corrector Method is the combination of the fourth-order Adams-Moulton Method and the fourth-order Adams-Bashforth Method.

At first, the Adams-Bashforth Formula is used to calculate the predicted value of the y_{k+1}.

Predict:

$$y_{k+1}^p = y_k + \frac{h}{24}(55f_k - 59f_{k-1} + 37f_{k-2} - 9f_{k-3}) \tag{2.12}$$

This predicted value is then used to calculate f_{k+1}.

Modify:

$$f_{k+1}^p = f(x_{k+1}, y_{k+1}^p) \tag{2.13}$$

The value of f_{k+1}^p is then substituted in the implicit Adams-Moulton Formula to correct the value of y_{k+1}.

Correct:

$$y_{k+1} = y_k + \frac{h}{24}(9f_{k+1}^p + 19f_k - 5f_{k-1} + f_{k-2}) \tag{2.14}$$

2.2.3 Stiff Equations and Implicit Methods

The concept of stiffness is related to stability and can be understood in the context of the following simple problem of a batch reactor. The first order reaction of a batch reactor is given below.

Consider the differential equation

$$\frac{dC_A}{dt} = -kC_A, \quad C_A(0) = C_{A0} \tag{2.15}$$

where C_A is the concentration of the reactant A. C_{A0} is the initial concentration and k is the rate constant.

Let us assume k to be 5.

The exact solution to this problem is given by:

$$C_A(t) = C_{A0}e^{-5t} \tag{2.16}$$

If we apply Euler's method to this problem, recursively, one can arrive at the following approximation to the exact solution (equation 2.16).

$$C_A|_n = C_{A0}(1 - 5h)^n \tag{2.17}$$

The limiting value of C_A is zero, which can be obtained as t tends to ∞. The value of C_A decreases very rapidly until t is equal to 0.2, and then slowly approaches zero. Therefore, we would expect to obtain sufficient accuracy with Euler's method using a relatively large h. However, if one looks at the exact solution and Euler's approximation using h equal to 0.001 (Figure 2.1), the approximation diverges from the original solution rapidly. In fact, the concentration goes below zero. This is because the quantity $(1 - 5h)^n$ is an approximation to e^{-5h} and is a good approximation for a very very small h. Even though this exponential term contributes virtually nothing to the solution after $t = 0.5$, Euler's method still requires that we approximate it with sufficient accuracy to maintain stability. This is the typical problem with stiff equations: the solution contains a component that contributes very little to the solution, but the usual methods require it be approximated accurately.

Stiff initial value problems were first encountered in the study of the motion of springs of varying stiffness, from which the problem derives its name. For linear ordinary differential equations, the stiffness of the system can be defined in terms of the stiffness ratio SR (Finlayson, 1980), given by:

$$SR = \frac{max\,|Re\,\lambda_i|}{min\,|Re\,\lambda_i|} \tag{2.18}$$

where, $Re\,\lambda_i$ is a real part of an eigenvalue λ_i of the Jacobian matrix $[\frac{df}{dy}]$ for the set of ordinary differential equations defined as $y' = f(y)$. Typically, SR of the order of 10 is considered to be not stiff; SR around 10^3 is stiff; and SR around 10^6 is very stiff. For nonlinear ordinary differential equations, the eigenvalues correspond to the eigenvalues of the Jacobian matrix at that

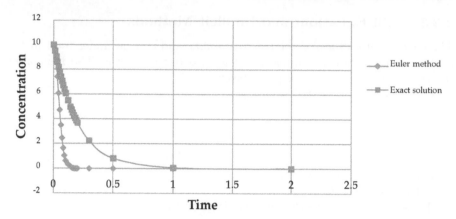

FIGURE 2.1
Comparison of Euler's method with exact solution

particular time step. The stiffness ratio then applies only to that particular time and may change as the integration proceeds.

The general approach to solving stiff equations is to use implicit methods. Historically, two chemical engineers, Curtis and Hirschfelder ([11]), proposed the first set of numerical formulas that are well-suited for stiff initial value problems by adopting:

$$y_{n+1} - y_n = hf_{n+1} \tag{2.19}$$

$$y_{n+2} - \frac{4}{3}y_{n+1} + \frac{1}{3}y_n = \frac{2}{3}f_{n+2} \tag{2.20}$$

Both schemes are implicit and belong to the well-known class of backward difference formula (BDF), the first one being the implicit Euler's scheme.

Consider the same differential equation presented earlier (Equation 2.15), and apply a backward Euler's method, which leads to the following solution.

$$C_A|_n = C_{A0}(1 + 5h)^{-n} \tag{2.21}$$

Now there is no unstable behavior regarding the size of h (Figure 2.2). Note that in the explicit Euler's method we were approximating the solution by a polynomial, and there exists no polynomial that can approximate the exponential term as x tends to ∞, hence, the instability. By using the implicit method, we have expressed the solution in the form of a rational function, which can go to zero, as t tends to ∞.

A detailed theory of stability and the different definitions of stability criteria (e.g., A-stable systems, stiffly stable systems, etc.) are beyond the scope of this book and readers are advised to look elsewhere for details ([9, 12, 13, 14, 15]). There are large numbers of stiff algorithms derived based

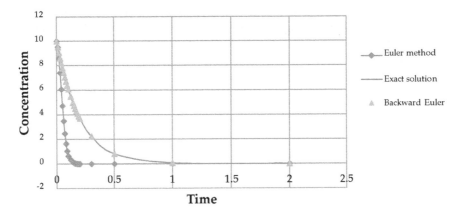

FIGURE 2.2
Comparison of Euler and backward Euler's methods with exact solution

on the above theory. The list of these algorithms are listed in Fatunla ([15])
and Diwekar [7].

One other way to solve stiff differential equations is to use orthogonal
collocation method. This method is described below.

2.3 Orthogonal Collocation Method

Consider a general nonlinear differential equation:

$$D(y) = f(x, y) \tag{2.22}$$

where $D(y)$ is the differential operator. In general, the solution of this equation
can be written as a combination of a series of known basis functions $\theta_i(x)$ and
unknown coefficients a_i given below.

$$y_s = \sum_{i=1}^{nn} a_i \theta_i(x) \tag{2.23}$$

We expect that for large values of nn ($nn \rightarrow \infty$), the solution in Equation 2.23 will approach the exact solution. The $\theta_i(x)$ denote the polynomial
basis functions of x and the coefficients a_i are obtained using the method of
weighted residuals (MWR), which involves substituting the approximate solution given by Equation 2.23 in Equation 2.22 to obtain a residual function
given below.

$$R(x, a) = D(y_s) - f(x, y_s) \tag{2.24}$$

The coefficients a_i are determined by minimizing the residual function $R(x, a)$ over the desired range of the independent variable x (for example, over $0 \leq x \leq 1$) along with a choice of the weighting function, $w(x)$. The residual minimization function may be written as follows:

$$\int_0^1 R(x, a)w(x)dx = 0 \tag{2.25}$$

Equation 2.25 describes the generalized form of all the methods of weighted residuals. The specific choice of the weighting function depends on the method used. The following list provides a few of these methods.

- Method of Least Squares

- Method of Galerkin [16]

- Method of Collocation [17], [18]

The least squares method was originally proposed by Gauss in 1795 in the context of parameter estimation. The weighting function in the least squares method is $dR(x, a)/da$ which results in:

$$\int_0^1 R(x, a)\frac{dR(x, a)}{da}dx = 0 \tag{2.26}$$

$$\text{Min} \int_0^1 (R(x, a))^2 dx \tag{2.27}$$

The least squares method can become very cumbersome for the solution of differential equations.

The Galerkin method, on the other hand, uses dy_s/da as the weighting function. The Galerkin method chooses an orthogonal class of functions and thus forces the residual to be zero by making it orthogonal to each member of a complete set (in the limit as $nn \rightarrow 0$). The Galerkin method is one of the best known approximate methods of weighted residuals.

The collocation method uses a delta function as a weighting function, given by

$$w(x) = \delta(x - x_k), k = 1, 2, \ldots, ncol \tag{2.28}$$

where the x_k are $ncol$ points between 0 and 1. This is equivalent to saying that the residual goes to zero at every collocation point.

$$R(a, x_k) = 0, k = 1, 2, \ldots, ncol \tag{2.29}$$

Now if the locations of the collocation points are the zeros of an orthogonal polynomial, the solution of Equation 2.23 approaches the Galerkin approximation [19], which has been shown to be the most accurate method of weighted residuals. The difference is that the Galerkin method is not suitable for machine computation while orthogonal collocation is easily programmable and is as accurate as the Galerkin method.

In the orthogonal collocation method, the collocation points are taken as roots of orthogonal polynomials. This procedure was first advanced by Lanczos [18] and was developed further for the solution of ordinary differential equations using the Chebyshev series. These applications were primarily for initial-value problems. In 1967, Villadsen and Stewart[20] made a major advance when they developed orthogonal collocation for boundary-value problems. They chose the functions to be sets of orthogonal polynomials that satisfied the boundary conditions while the roots of the polynomials gave the collocation points. Thus, the choice of the collocation points is no longer arbitrary and the lower-order collocation results are more dependable. A major simplification is that the solution can be derived in terms of the coefficients in the function and the values of the solution at the collocation points. The whole problem is then reduced to a set of matrix equations, which are easily generated and solved numerically [21].

In orthogonal collocation, the trial function given in Equation 2.23 is written in terms of linear combinations of orthogonal polynomials $P_m(x)$ of the order 1 to $m + 1$, with P_0 as the starting point.

$$y_s = \sum_{i=1}^{m} c_i P_{m-1}(x) \tag{2.30}$$

There are a number of different kinds of orthogonal polynomials one can use, including continuous polynomials like Lagrange or Legendre polynomials [22] ,[21], and discrete ones, such as Hahn's polynomial [23]. The orthogonality property allows one to obtain the roots of the polynomial x_i, $i = 1, 2, \ldots, m - 1$. Since orthogonal polynomials are also formed by linear combination of x or x^2 (for simplicity we can take the example of polynomials in x), Equation 2.30 can be rewritten at each collocation point in terms of new coefficients d_i

$$y_j = \sum_{i=1}^{m} d_i x_{i-1}; \qquad j = 1, 2, \ldots, m - 1 \tag{2.31}$$

Differentiating the above equation results in:

$$\frac{dy}{dx}\Big|_{x_j} = \sum_{i=1}^{m} d_i \frac{dx_{i-1}}{dx}\Big|_{x_j}; \qquad j = 1, 2, \ldots, m - 1 \tag{2.32}$$

$$\frac{d^2y}{dx^2}\Big|_{x_j} = \sum_{i=1}^{m} d_i \frac{d^2x_{i-1}}{dx^2}\Big|_{x_j}; \qquad j = 1, 2, \ldots, m - 1 \tag{2.33}$$

Substituting the value of d_i from Equation 2.31 in the above equations:

$$\frac{dy}{dx}\Big|_{x_j} = \sum_{i=1}^{m} A_{ji}y_i; \qquad j = 1, 2, \ldots, m - 1 \tag{2.34}$$

$$\frac{d^2y}{dx^2}\Big|_{x_j} = \sum_{i=1}^{m} B_{ji}y_i; \qquad j = 1, 2, \ldots, m - 1 \tag{2.35}$$

To evaluate the integral, a simple quadrature formula can be used.

$$\int_{o}^{1} y\,dx = \sum_{i=1}^{m} W_i f(x_i, y_i) \tag{2.36}$$

where

$$[A_{ji}] = [\tfrac{dx_{i-1}}{dx}|_{x_j}][x_{i-1}]^{-1}$$

$$[B_{ji}] = [\tfrac{d^2 x_{i-1}}{dx^2}|_{x_j}][x_{i-1}]^{-1}$$

$$[W_i] = [f(x_j, y_j)][x_{i-1}]^{-1}$$

From Equations 2.34 and 2.35, one can easily see that the differential operator can be replaced by a linear combination of the solution y at the collocation points $j = 1, 2, \ldots, m - 1$ and, hence, the ordinary differential equations are reduced to a set of algebraic equations as given below.

$$\sum_{i=1}^{m} D_{ji}y_i = f(x_j, y_j); \qquad j = 1, 2, \ldots, m - 1 \tag{2.37}$$

With this technique, the partial differential equations can be reduced to a set of ordinary differential equations (or a set of algebraic equations). The following example will illustrate the orthogonal collocation technique for the initial-value problem.

Example 2.1: Solve the differential Equation 2.15 using the orthogonal collocation technique for different numbers of collocation points and compare it with the analytical solution in the range 0 to 0.125.

$$\frac{dC_A}{dt} = -kC_A, \quad C_A(0) = C_{A0}$$

For this purpose, use the shifted Legendre polynomials whose roots and matrices are provided below.

Roots and Matrices of Shifted Legendre Polynomials (Finlayson, 1972)

$$N = 1 \quad x = \begin{bmatrix} 0 \\ 0.5000 \\ 1 \end{bmatrix} \quad W = \begin{bmatrix} \frac{1}{6} \\ \frac{2}{3} \\ \frac{1}{6} \end{bmatrix}$$

$$A = \begin{bmatrix} -3 & 4 & -1 \\ -1 & 0 & 1 \\ 1 & -4 & 3 \end{bmatrix} \quad B = \begin{bmatrix} 4 & -8 & 4 \\ 4 & -8 & 4 \\ 4 & -8 & 4 \end{bmatrix}$$

Solution: The analytical solution of the above differential equation is given by

$$C_A(t) = C_{A0}e^{-5t}$$

Since we want to integrate from 0 to 0.125 and the roots are 0, 0.5, 1.0, we need to scale the differential equation. Let us substitute $\tau = t/8.0$. Then the differential equation changes to:

$$\frac{dC_A}{d\tau} = -k/8.0 C_A, \quad C_A(0) = C_{A0}$$

In the collocation method the solution is obtained at collocation points ($\tau_1 = 0; \tau_2 = 0.5; \tau_3 = 1.0$). From the above analytical expression the values of C_A at the three collocation points are given by

$$C_{A_1} = 10; \quad C_{A_2} = 7.3; \quad C_{A_3} = 5.4$$

Now we will solve the problem using the collocation method. From Equation 2.34 the following equation can be derived

$$\sum_{i=1}^{ncol+2} A_{ij}C_{A_j} = -5/8 C_{A_j}; \qquad j = 2, \dots, ncol + 2$$

Substituting the values of t_j in the above equations results in the following equations in t_j.

$$
\begin{aligned}
-3C_{A_1} + 4C_{A_2} - 1C_{A_3} &= -5/8 C_{A_1} \\
-1C_{A_1} + 0C_{A_2} + 1C_{A_3} &= -5/8 C_{A_2} \\
1C_{A_1} - C_{A_2} + 3C_{A_3} &= -5/8 C_{A_3}
\end{aligned}
$$

Substituting $C_{A_1} = 10$, the simultaneous solution of the two of the above three equations provides

$$C_{A_1} = 10; \quad C_{A_2} = 7.4; \quad C_{A_3} = 5.4$$

which is close to the analytical solution. If we want to integrate further, with the same number of collocation points, we can use the finite element collocation as shown in Figure 2.3.

While most of the batch unit operation models involve ordinary differential equations some unit operations like batch adsorption column encounters partial differential equations, orthogonal collocation method can be used to reduce set of partial differential equations to ordinary differential equations.

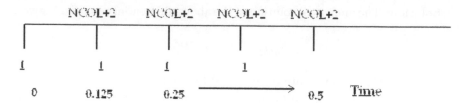

FIGURE 2.3
Finite element collocation

2.4 Summary

Numerical integration techniques are necessary in modeling and simulation of batch and bio processing. In this chapter we described error and stability criteria for numerical techniques. Various numerical techniques for solution of stiff and non-stiff problems are discussed. These methods include one-step and multi-step explicit methods for non-stiff and implicit methods for stiff systems, and orthogonal collocation method for ordinary as well as partial differential equations. These methods are an integral part of some of the packages like MATLAB. However, it is important to know the theory so that appropriate method for simulation can be chosen.

Notations

C_A	concentration of A [mol/vol]
C_{A0}	concentration of A at time=0 [mol/vol]
h	step size
K	reaction constant [time^{-1}]
$O(h^p)$	p-th order method
m	order of approximate difference equation
q	order of original differential equation
SR	stiffness ratio
t	integration variable, time [hr]
x	integration variable
y	state variable

Greek Letters:

ϵ	rounding error
η	difference approximation
λ	eigenvalue

3

Batch Reactors

CONTENTS

Batch reactors are commonly used in pharmaceutical as well as speciality chemical industry where quality of the product is given enormous importance. In case of bio reactions such as fermentation, a variant of batch reactor called a fed-batch reactor is commonly used. Figure 3.1 shows a schematic of these two reactors. As seen in the figure, in a batch reactor, the reactor is filled with reactants and allowed to react for a set time. While in a fed-batch reactor apart from starting with a reactor filled with reactants, you keep adding feed continuously until you reach the capacity.

Many batch reactors deal with constant-density reaction systems. This includes most liquid-phase reactions as well as all gas phase reactions occuring in a constant-volume bomb. This is a constant volume batch reactor and is the focus of this chapter. Much of this chapter is derived from Levenspiel[24].

3.1 Classification of Reactions and Reaction Kinetics

There are many ways to classify reactions. Depending on the number and types of phases involved, the reactions can be classified as homogenous or heterogenous. Homogeneous reactions take place in one phase only while heterogenous reactions involve more than one phase. This classification is not so clear in case of biological reactions involving enzymes. As enzymes are highly complicated large-molecular-weight proteins of colloidal sizes 10-100 nm. Enzyme containing solution represents a grey area between heterogenous and homogenous systems. Reactions can also be classified as catalytic or non-catalytic reac-

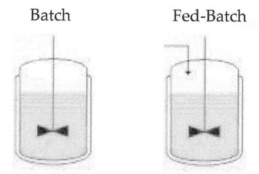

Batch   Fed-Batch

FIGURE 3.1
Schematic of batch and fed-batch reactors

tions. The next section describes various reactions classified based on their kinetics.

In a constant volume system the measure of reaction rate (kinetics) for component i, r_i can be written as

$$r_i = \frac{1}{V}\frac{dN_i}{dt} = \frac{dC_i}{dt} \tag{3.1}$$

where V is volume of fluid, N_i number of moles of component i, and C_i is the molar concentration of component i.

Order of Reaction

The reaction kinetics or rate of reaction can be described based on concentration of reactants $(A, B, C \ldots)$ as below.

$$-r_A = -\frac{dC_A}{dt} = kC_A^a C_B^b C_C^c \ldots, \quad a + b + c + \cdots = n \tag{3.2}$$

where k is the reaction constant and the reaction is said to be

a^{th} order with respect to A

b^{th} order with respect to B

n^{th} order overall

Arrehenius Law

The rate constant in a chemical reaction k consists of two parts as given below

$$k = Ae^{-E/RT} \tag{3.3}$$

where A is called the frequency or pre-exponential factor, E is the activation energy of reaction, R is the gas constant, and T is reaction temperature. This equation is called the Arrehenius law and is a good approximation to the temperature dependency of reaction.

This temperature dependency is exploited in optimal control problems of batch reactor where optimal temperature profile is obtained by either maximizing conversion, yield, profit, or minimizing batch time for the reaction. One of the earliest works on optimal control of batch reactor was presented by Denbigh[25] where he maximized the yield. The review paper by Srinivasan et al.[26] describes various optimization and optimal control problems in batch processing and provides examples of semi-batch and fed-batch reactor optimal control.

3.2 Irreversible Reactions

First-Order Reactions

Consider a uni-molecular reaction given below.

$$A \xrightarrow{k} products \tag{3.4}$$

The first-order rate equation for this reaction can be written as

$$
\begin{aligned}
-r_A &= r_{product} \\
-r_A &= -\frac{dC_A}{dt} = kC_A
\end{aligned}
\tag{3.5}
$$

Transforming and integrating Equation 3.5 results in

$$-\int_{C_{A0}}^{C_A} \frac{dC_A}{C_A} = k \int_0^t dt \tag{3.6}$$

$$-\ln \frac{C_A}{C_{A0}} = kt \tag{3.7}$$

These equations can be written in terms of conversion X_A as follows

$$-\int_0^{X_A} \frac{d(C_{A0}(1-X_A))}{(C_{A0}(1-X_A))} = k \int_0^t dt \tag{3.8}$$

$$-\ln(1-X_A) = kt \tag{3.9}$$

Second-Order Reactions

Consider the reaction and corresponding reaction rate given below.

$$A + B \xrightarrow{k} products \tag{3.10}$$

$$-\frac{dC_A}{dt} = kC_A C_B \tag{3.11}$$

If the conversion of A is given by X_A, given the Equation 3.11, the moles of B converted are the same as the moles of A which is $C_{A0}X_A$ resulting in the following rate equation in terms of conversion.

$$
\begin{aligned}
-r_A &= -\frac{d(C_{A0}(1 - X_A))}{dt} \\
&= k(C_{A0} - C_{A0}X_A)(C_{B0} - C_{A0}X_A)
\end{aligned}
\tag{3.12}
$$

Integration of this equation results in

$$\ln\frac{(C_B C_{A0})}{(C_A C_{B0})} = -\frac{(1 - X_B)}{(1 - X_A)}dt = kt(C_{B0} - C_{A0}) \tag{3.13}$$

With an equal concentration of A and B, Equation 3.13 can be written as:

$$\frac{1}{C_A} - \frac{1}{C_{A0}} = kt \tag{3.14}$$

Figure 3.2 shows the transient concentration in a batch reactor for a first-order and second-order reactions with the molecule A where the reaction constant $k = 0.5 \ time^{-1}$ and $C_{A0} = 3 \ moles/volume$. As can be seen the first-order reaction concentration decreases slower initially as compared to second-order reaction as the concentration of reactant is higher. After time $t = 4$, the trend reverses as the concentration reduces beyond 0.5.

n-th Order Reactions

When the order of reaction is not known, we fit the following equation to the $n - th$ order reaction where $n > 1$.

$$-r_A = -\frac{dC_A}{dt} = kC_A{}^n \tag{3.15}$$

Separating and integrating the above equation results in

$$C_A{}^{n-1} - C_{A0}{}^{n-1} = (n - 1)kt \tag{3.16}$$

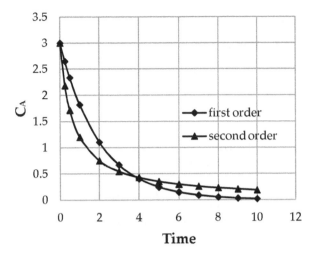

FIGURE 3.2
Concentration profiles for the first-order and second-order reactions

Reactions of Shifting Order

In some reactions, the order of reaction changes with time. Biological reactions like fermentation discussed in the autocatalytic reaction section shows such phenomena. For example, consider again the uni-molecular reaction with the following rate equation.

$$A \rightarrow R \qquad (3.17)$$

$$-r_A = -\frac{dC_A}{dt} = \frac{k_1 C_A}{1 + k_2 C_A} \qquad (3.18)$$

This reaction is zeroth order for high concentration of C_A with rate constant k_1/k_2 and is first-order with rate constant k_1 for low concentration of C_A.

By using separation of variables, the rate equation can be integrated results in

$$\frac{\ln\left(\frac{C_A}{C_{A0}}\right)}{C_{A0} - C_A} = -k_2 + \frac{k_1 t}{C_{A0} - C_A} \qquad (3.19)$$

3.3   Reversible Reactions

Some reactions cannot go to completion as there are reversible reactions. A simplest case of such a reaction is given below

$$A \underset{\longleftarrow}{\overset{\longrightarrow}{k_1, k_2}} R \tag{3.20}$$

where the reaction constant for the forward reaction is k_1 and the backward reaction k_2. Then the reaction rate for this reaction can be written as

$$\frac{dC_R}{dt} \quad = \quad -\frac{dC_A}{dt} = k_1 C_A - k_2 C_R \tag{3.21}$$

$$= \quad k_1(C_{A0} - C_{A0}X_A) - k_2(C_{B0} + C_{A0}X_A) \tag{3.22}$$

We know that the equilibrium rate of reaction is zero with equilibrium conversion X_{Ae}. Equating Equation 3.22 to zero

$$X_{Ae} \quad = \quad \frac{k_1 C_{A0} - k_2(C_{B0})}{C_{A0}(k_1 + k_2)} \tag{3.23}$$

Integrating Equation 3.22 and substituting values of X_{Ae} from Equation 3.23 results in

$$-\ln\left(1 - \frac{X_A}{X_{Ae}}\right) \quad = \quad \frac{C_{R0}/C_{A0} + 1}{C_{R0}/C_{A0} + X_{Ae}} k_1 t \tag{3.24}$$

For an order other than one or two, integrating the rate equation requires numerical integration techniques.

3.4   Reactions in Parallel

The simplest case of reactions in parallel can be written as, decomposition of component A by two competing paths (both elementary reaction) given by

$$A \overset{\longrightarrow}{k_1} R$$
$$A \overset{\longrightarrow}{k_2} S \tag{3.25}$$

The reaction rates for all three components are given by

$$-r_A \quad = \quad -\frac{dC_A}{dt} = k_1 C_A + k_2 C_A \tag{3.26}$$

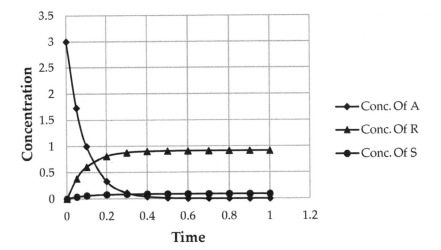

FIGURE 3.3

Concentration profiles for the parallel reactions

$$r_R \quad = \quad \frac{dC_R}{dt} \quad = \quad k_1 C_A \tag{3.27}$$

$$r_S \quad = \quad \frac{dC_S}{dt} \quad = \quad k_2 C_A \tag{3.28}$$

Integrating Equation 3.26 results in

$$-\ln \frac{C_A}{C_{A0}} \quad = \quad (k_1 + k_2)t \tag{3.29}$$

From Equations 3.27 and 3.28, we can write

$$\frac{r_R}{r_S} \quad = \quad \frac{dC_R}{dC_S} \quad = \quad \frac{k_1}{k_2} \tag{3.30}$$

This equation when integrated results in

$$\frac{C_R - C_{R0}}{C_S - C_{S0}} \quad = \quad \frac{k_1}{k_2} \tag{3.31}$$

Figure 3.3 shows the concentration profile for the parallel reactions where $k_1 = 10 \ time^{-1}$, $k_2 = 1 \ time^{-1}$, and $C_{A0} = 3 \ moles/volume$.

3.5 Reactions in Series

Again here we consider the simplest reactions in a series given below.

$$A \underset{\rightarrow}{k_1} R \underset{\rightarrow}{k_2} S \tag{3.32}$$

The rate equation for this reaction assuming first-order kinetics can be written as

$$-r_A = -\frac{dC_A}{dt} = k_1 C_A \tag{3.33}$$

$$r_R = \frac{dC_R}{dt} = k_1 C_A - k_2 C_R \tag{3.34}$$

$$r_S = \frac{dC_S}{dt} = k_2 C_R \tag{3.35}$$

Integrating Equation 3.33 results in

$$-\ln \frac{C_A}{C_{A0}} = kt \tag{3.36}$$

$$C_A = C_{A0} e^{-k_1 t} \tag{3.37}$$

Substituting in Equation 3.34 and integrating

$$\frac{dC_R}{dt} = k_1 C_A 0 e^{-k_1 t} - k_2 C_R \tag{3.38}$$

$$C_R = C_{A0} k_1 \left(\frac{e^{-k_1 t}}{k_2 - k_1} - \frac{e^{-k_2 t}}{k_2 - k_1} \right) \tag{3.39}$$

We know from the reaction Equation 3.32 that the total number of moles remains constant, therefore,

$$C_{A0} = C_A + C_R + C_S \tag{3.40}$$

Substituting values of CA and C_R from equations 3.37 and 3.39 results in

$$C_S = C_{A0} \left(1 + \frac{k_2}{k_1 - k_2} e^{-k_1 t} - \frac{k_1}{k_2 - k_1} e^{-k_2 t} \right) \tag{3.41}$$

Figure 3.4 shows the concentration profile for the series of reactions where $k_1 = 10 \ time^{-1}$, $k_2 = 1 \ time^{-1}$, and $C_{A0} = 3 \ moles/volume$. From this figure

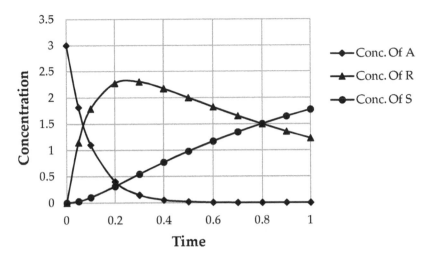

FIGURE 3.4
Concentration profiles for the reactions in series

it is clear that the product R shows a maximum value when $dC_R/dt = 0$ (please refer to chapter on optimization, NLP section). Thus the time at which the maximum concentration of R occurs is given by

$$t_{max} \quad = \quad \frac{\ln k_2/k_1}{k_2 - k_1} \tag{3.42}$$

which for this case shown in the figure is 0.256.

3.6   Autocatalytic Reactions

A reaction in which one of the products of reaction acts as a catalyst is called an autocatalytic reaction. In a n-th order in a batch reactor, the rate of product formation or disappearance of reactant is high initially as concentration of the reactant is high, and reaction slows down as the reactant disappears. However, in an autocatalytic reaction, the rate is low initially as little product is present, the rate increases as more and more product gets formed and then drops again because reactant is consumed.

The simplest case of such a reaction is given below.

$$A + R \xrightarrow{k_1} R + R \tag{3.43}$$

For this reaction, a reaction rate based on first-order kinetics can be written as

$$-\frac{dC_A}{dt} \quad = \quad k_1 C_A C_R \tag{3.44}$$

We know from the reaction Equation 3.43 that the total number of moles remains constant, therefore,

$$C_0 = C_{A0} + C_{R0} \quad = \quad C_A + C_R \tag{3.45}$$

Substituting value of C_R from Equation 3.45 into Equation 3.44 and integrating results in

$$\ln \frac{[C_{A0}(C_0 - C_A)]}{[C_A(C_0 - C_{A0})]} \quad = \quad C_0 kt \tag{3.46}$$

Figure 3.5a shows the concentration profile for this reaction with $k = 0.5\ time^{-1}$, $C_{A0} = 3\ moles/volume$, and $C_{R0} = 1.0\ moles/volume$ and Figure 3.5b shows the reaction rate. As can be seen the reaction rate is parabolic, it is slower initially and goes through a maximum at $C_A = C_R$ and again decreases. For an autocatalytic reactor, some product (R here) should be present to start the reaction.

Reactions in which living cells are used to produce chemicals are autocatalytic in nature. The rate of reaction is proportional with the biomass concentration, and biomass is one of the products of the reaction. The most commonly found examples of autocatalytic reaction are the broad class of fermentation reactions which result from the reaction of microorganisms on an organic feed. Rate expressions for bioreactions are grouped into expressions that can be used for balanced growth and expressions that are suitable to simulate unbalanced growth. At first it might seem absurd that one has to work with two sets of rate expressions, but the difference between the two lies only in the amount of detail included in the model. Conversion of substrates to products in a batch reactor is of course in principle a transient process, but for all practical purposes one may regard growth to be balanced in the batch reactor[27].

The most frequently used rate expression for balanced growth was proposed by Jacques Monod in 1942[28] and is known as the Monod rate expression.

Cellular growth (X) rate

$$\frac{dX}{dt} \quad = \quad \mu X \tag{3.47}$$

where μ is the specific growth rate given by Monod rate expression below.

Monod rate expression

$$\mu \quad = \quad \mu_{max} \frac{S}{K_s + S} \tag{3.48}$$

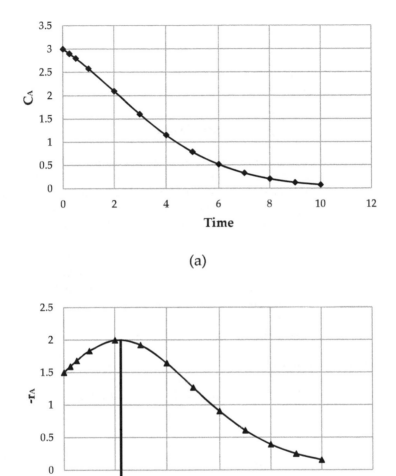

(a)

(b)

FIGURE 3.5
Concentration profiles for the autocatalytic reactions

where μ_{max} is the maximum specific growth rate of the microorganisms expressed in units of $time^{-1}$, and K_s is the half-saturation coefficient expressed in mass per unit volume.

Substrate (S) utilization

$$-\frac{dS}{dt} = \frac{\mu_{max}}{Y_{X/S}}\frac{SX}{K_s + S} \tag{3.49}$$

where $Y_{X/S}$ is the yield coefficient given by $\frac{\Delta X}{\Delta S}$.

Product (P) formation

$$\frac{dP}{dt} = \mu_{max}Y_{X/S}\frac{SX}{K_s + S} \tag{3.50}$$

Figure 3.6 shows dynamics of cellular growth and substrate for a particular species in the batch reactor (above equations) and a fed-batch reactor. As can be seen, these reactions provide better yield in fed-batch reactors. As stated earlier, in a fed-batch reactor, feed is added continuously. Determining optimal feeding profile is a commonly addressed optimal control problem for fed-batch reactors[26].

The Monod equation has the same form as the Michaelis-Menten[29] equation for enzyme kinetics, but differs in that it is empirical while the latter is based on theoretical considerations. For the simple enzyme reaction given below, the Michaelis-Menten equations are provided below.

$$E + S \quad \underset{\longleftrightarrow}{k_1, k_{-1}} \quad ES \underset{\rightarrow}{k_2} E + P \tag{3.51}$$

Michaelis-Menten model

$$-\frac{dC_{ES}}{dt} = k_1 C_E C_S - k_{-1}C_{ES} - k_2 C_{ES} \tag{3.52}$$

$$-\frac{dP}{dt} = \frac{k_2 C_E C_S}{\frac{k_{-1}+k_2}{k_1} + C_S} \tag{3.53}$$

$$= \frac{V_{max}}{\frac{C_S}{K_m + C_S}} \tag{3.54}$$

As can be seen Equation 3.54 is a Monod rate expression.

The following example presents manufacturing of tequila in a batch reactor.

Example 3.1: Recently Harrera et al. [30] presented a model for tequila fermentation based on experimental data. This model is described below. If the fermentation is carried out in a batch reactor, find the concentration profiles for substrate, biomass, and product from time, $t = 0$ to $t = 100$ hrs.

The fermentation reaction is given below.

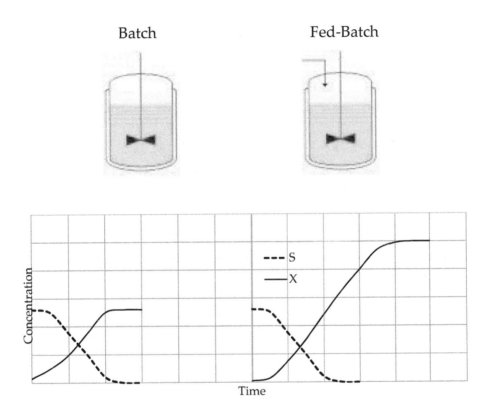

FIGURE 3.6
Concentration profiles for the Monod kinetics in a batch and a fed-batch reactor

TABLE 3.1

Parameters for the kinetic model

Parameter	Value
μ_{max}	0.5372
K_s	20.4131
K_1	0.0089
K_p	0.02669
Y_{xs}	17.98
m_s	0.0053
α	8.7053
S_0	90
X_0	0.5
P_0	2.5

$$S + X \rightarrow X + X + P + CO_2 \tag{3.55}$$

where S is the substrate, X is biomass, and P is ethanol.

Kinetic model for tequila fermentation is given below

$$\frac{dX}{dt} = \mu X \tag{3.56}$$

where μ is the specific growth rate given by modified Monod rate expression below.

Modified Monod rate expression

$$\mu = \mu_{max} \frac{S}{K_s + S + K_1 S^2} (1 - K_p P) \tag{3.57}$$

where μ_{max} is the maximum specific growth rate of the microorganisms expressed in terms of $time^{-1}$, K_s is the half-saturation coefficient expressed in mass per unit volume, and K_1 and K_p are inhibition parameters for the substrate and product respectively.

Substrate (S) utilization

$$-\frac{dS}{dt} = -\mu(Y_{xs} - Y_{ps})X - m_s X \tag{3.58}$$

where Y_{xs} and Y_{ps} is the yield coefficient given by $Y_{xs} = \frac{\Delta X}{\Delta S}$, $Y_{ps} = \frac{\Delta P}{\Delta S}$, and m_s (g/L) is the cell maintenance coefficient.

Product (P) formation

$$\frac{dP}{dt} = \alpha \mu X \tag{3.59}$$

where α is a term associated with growth

Table 3.1 provides the parameters for the kinetic model given above.

Solution: Given equations for the batch reactor fermentation kinetics are difficult to integrate analytically, we looked at numerical integration techniques to solve

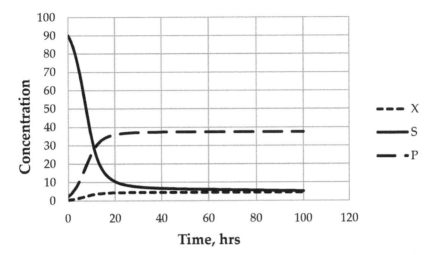

FIGURE 3.7
Concentration profiles for the tequila fermentation in a batch reactor

this problem. To provide better accuracy, we used the predictor-corrector method to solve this problem.

Figure 3.7 shows the concentration profiles for all the three components obtained by numerically integrating equations 3.56 to 3.59.

3.7 Biodiesel Production: A Case Study

Amongst all the biofuels, biodiesel is considered one of the best biofuel candidates. Biodiesel is superior to diesel oil in terms of sulfur and aromatic content, and it is environmentally safe, non-toxic, and biodegradable. Biodiesel or fatty acid methyl esters are products of the transesterification process. This process is achieved by the reaction of triglycerides which are contained in oils like the soybean oil, with an excess of alcohol (i.e., methanol) in the presence of an acid and alkaline catalyst (i.e., sodium hydroxide). The reaction consists of three stepwise and reversible reactions where in the first step the triglycerides with one molecule of methanol are converted into diglycerides, then, the diglycerides are converted into monoglycerides and this last one into one molecule of glycerol. In addition, in each step a methyl ester (i.e., $R_i COOCH_3$) is produced resulting in three molecules of methyl ester from one molecule of triglycerides. These reaction steps presented by Noureddini and Zhu, 1997 [31] are shown in Figure 3.8. These reactions are carried out in a batch reactor and hence this case study is included in this chapter. In this section, we will present the batch reactor design and optimal control problems for biodiesel production derived from Benavides and Diwekar (2012)[32].

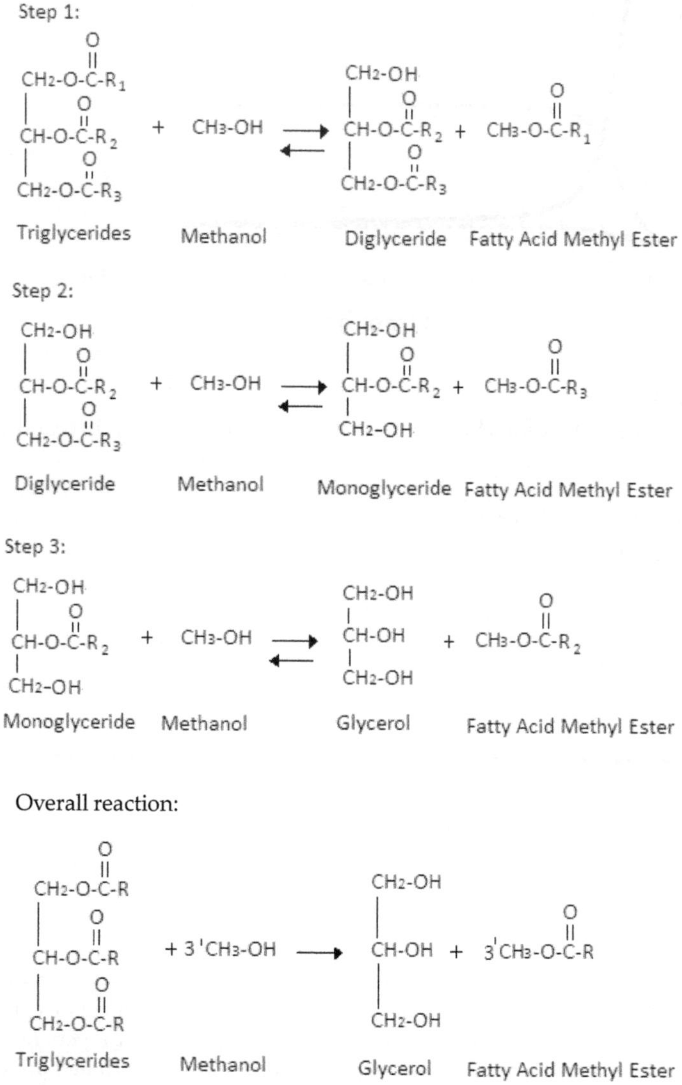

Step 1:

Triglycerides   Methanol   Diglyceride   Fatty Acid Methyl Ester

Step 2:

Diglyceride   Methanol   Monoglyceride Fatty Acid Methyl Ester

Step 3:

Monoglyceride   Methanol   Glycerol   Fatty Acid Methyl Ester

Overall reaction:

Triglycerides   Methanol   Glycerol   Fatty Acid Methyl Ester

FIGURE 3.8
Biodiesel transesterification reactions

TABLE 3.2
Values of a_i and b_i

a_1	a_2	a_3	a_4	a_5	a_6
3.92e7	5.77e5	5.88e12	098e10	5.35e3	2.15e4
b_1	b_2	b_3	b_4	b_5	b_6
6614.83	4997.98	9993.96	7366.64	3231.18	4824.87

Mathematical Model for Transesterification of Oil

The kinetic model based on the mechanism of transesterification for biodiesel reaction shown in Figure 3.8 presented below.

$$TG + CH_3OH \quad \overset{k_1, k_2}{\underset{}{\longleftrightarrow}} \quad DG + R_1COOCH_3$$
$$DG + CH_3OH \quad \overset{k_3, k_4}{\underset{}{\longleftrightarrow}} \quad MG + R_1COOCH_3$$
$$MG + CH_3OH \quad \overset{k_5, k_6}{\underset{}{\longleftrightarrow}} \quad GL + R_1COOCH_3 \qquad (3.60)$$

where k_1 to k_6 are reaction constants.

Overall reaction:

$$TG + 3\,CH_3OH \quad \longleftrightarrow \quad GL + 3\,R_1COOCH_3 \qquad (3.61)$$

where TG, DG, MG, R_1COOCH_3, CH_3OH, and GL are triglycerides, diglycerides, monoglycerides, methyl ester (i.e., biodiesel), methanol, and glycerol respectively.

$$k_i = a_i \, exp(-\frac{b_i}{T}) \qquad (3.62)$$

where a_i and b_i are Arrehenius equation constants and are presented in Table 3.2.

The mathematical model for the production of biodiesel in a batch reactor is governed by the following ordinary differential equations 3.63 to 3.68 derived from the mass balance of the batch reactor [31].

$$\frac{dC_{TG}}{dt} = -k_1 C_{TG} C_A + k_2 C_{DG} C_E \qquad (3.63)$$

$$\frac{dC_{DG}}{dt} = k_1 C_{TG} C_A - k_2 C_{DG} C_E - k_3 C_{DG} C_A + k_4 C_{MG} C_E \qquad (3.64)$$

$$\frac{dC_{MG}}{dt} = k_3 C_{DG} C_A - k_4 C_{MG} C_E - k_5 C_{MG} C_A + k_6 C_{GL} C_E \qquad (3.65)$$

$$\frac{dC_E}{dt} = k_1 C_{TG} C_A - k_2 C_{DG} C_E + k_3 C_{DG} C_A - k_4 C_{MG} C_E$$
$$+ k_5 C_{MG} C_A - k_6 C_{GL} C_E \qquad (3.66)$$

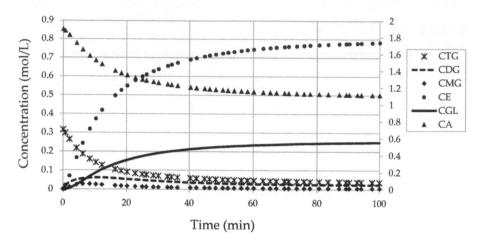

FIGURE 3.9
Concentration profiles for the transesterification reactions in a batch reactor
at constant temperature

$$\frac{dC_A}{dt} = -\frac{dC_E}{dt} \tag{3.67}$$

$$\frac{dC_{GL}}{dt} = k_5 C_{MG} C_A - k_6 C_{GL} C_E \tag{3.68}$$

where $C_{TG}, C_{DG}, C_{MG}, C_E, C_A,$ and C_{GL} are the concentration of each component.

It has been shown in the literature that the optimal constant temperatures for
transesterification reaction are between 323K to 333K at atmospheric pressure [33].
The differential equations are difficult to integrate analytically so a numerical integration technique based on the Runge-Kutta method is utilized. Figure 3.9 shows
the concentration profile for the six components for the case of 323K of reaction temperature. It can be seen as the triglycerides and methanol are consumed biodiesel is
produced.

The Optimal Control Problem

The optimal control problem encountered in biodiesel production in a batch reactor
can be classified into three categories depending on the objective function as given
below. The decision variable or the control variable is the temperature profile for
the reactor.

1. Maximum concentration problem: In this problem the final concentration
    of methyl ester is maximized by fixing batch time in the reactor.

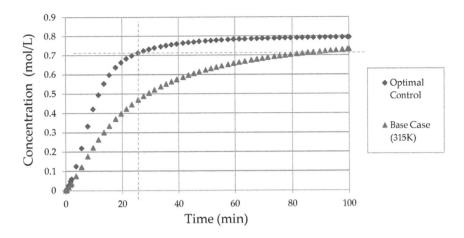

FIGURE 3.10
Concentration profiles for the methyl ester

2. Minimum time problem: In this problem batch time is minimized for a specific final concentration of methyl ester.
3. Maximum profit problem: In this problem neither batch time nor final concentration is fixed but a profit function involving batch time and final concentration is maximized.

Here we present the solution of the maximum concentration problem below. All three problems discussed above are solved in Benavides and Diwekar (2013)[34]. Interested readers are referred to this paper.

There are numbers of different methods to solve an optimal control problem as discussed in the chapter on optimization and optimal control. Benavides and Diwekar (2011, 2013)[32, 34] used the maximum principle to solve the maximum concentration problem where the batch time was fixed at 100 minutes. Figure 3.10 shows the concentration profile for the base case versus the profile obtained using optimal temperature profile shown in Figure 3.11. It can be seen that the optimal concentration for methyl ester is found to be 0.7944 mol/L, while at constant temperature, the maximum concentration is 0.7324 mol/L (8.46% gain). Alternatively, if we fix the concentration at 0.7324 mol/L, the reaction time needed would be 69.5% less than it was at the beginning.

3.8 Summary

This chapter presented analysis of batch reactors for chemical as well as biological reactions. Various types of reactions such as first-order, second-order, n-th order, shifting order, reversible, irreversible, and autocatalytic are discussed and presented in simplest form. Monod and Michaelis-Menten models are discussed in the context

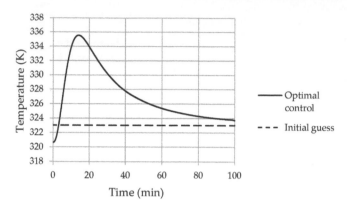

FIGURE 3.11
Optimal temperature profile

of bio reactions. A case study of biodiesel production in a batch reactor is presented. The case study discussed batch reactor simulation as well as solution of the maximum concentration optimal control problem.

Notations

$A, a_1, a_2, \ldots, a_6$	Arrehenius constant/frequency/pre-exponential factor $[\text{time}^{-1}]$
$b_1, b_2, \ldots, b_6$	Arrehenius exponential factor $[1/^{0}\text{K}]$
C_i	concentration of component i $[\text{mol/vol}]$
C_{i0}	concentration of component i at time=0 $[\text{mol/length}^3]$
E	activation energy of reaction $[\text{cal}]$
$k, k_{-1}, k_1, k_2, \ldots, k_6$	reaction constants $[\text{time}^{-1}]$
K_1	inhibition parameter for the substrate
K_p	inhibition parameter for the product
K_s	half-saturation coefficient $[\text{mass/length}^3]$
m_s	is the cell maintenance coefficient $[\text{gms/liter}]$
N_i	number of moles of component i $[\text{mol}]$
P	product concentration $[\text{mass/vol}]$
R	gas constant,1.98719 $[\text{cal/mol}/^{0}\text{K}]$
S	substrate concentration $[\text{mass/vol}]$
r_i	rate of reaction of production of component i $[\text{mol}]$
t	integration variable, time $[\text{time}]$
T	temperature, $[^{0}\text{K}]$
V	volume of fluid $[\text{vol}]$
X	cellular concentration $[\text{mass/length}^3]$
X_i	conversion of component i
X_{ie}	equilibrium conversion of component i

Y_{xs} yield coefficient given by $Y_{xs} = \frac{\triangle X}{\triangle S}$

Y_{ps} yield coefficient given by $Y_{ps} = \frac{\triangle P}{\triangle S}$

Greek Letters:

α term associated with growth Equation 3.59

μ specific growth rate given by Monod rate expression

μ_{max} maximum specific growth rate of the microorganisms [time $^{-1}$]

4

Batch Distillation

CONTENTS

Batch distillation[1] is the oldest separation process and the most widely used unit operation in pharmaceutical and specialty chemical industries. The most outstanding feature of batch distillation is its flexibility. This flexibility allows one to deal with uncertainties in feed stock or product specification.

In the distillation process, it is assumed that the vapor formed within a short period is in thermodynamic equilibrium with the liquid. Hence, the vapor composition y is related to the liquid composition x by an equilibrium relation of the functional form $y = f(x)$. The exact relationship for a particular mixture may be obtained from a thermodynamic analysis and is also dependent upon temperature and pressure. Figure 4.1 shows an example equilibrium curve for a system consisting of CS_2 and CCl_4 at 1 atmosphere pressure.

Simple distillation is the simplest form of batch distillation. In this type of distillation, a still is initially filled with a feed mixture, which evaporates after heating and leaves the still in the vapor form. This vapor which is richer in the more volatile component is collected in the condenser at the top. This simple distillation is often referred to as Rayleigh distillation because of Rayleigh's pioneering theoretical work [35]. The concept of reflux and the use of accessories such as plates and packing materials to increase the mass transfer converts this simple still into a batch distillation column as shown in Figure 4.2a. Because this batch column essentially performs the rectifying operation, it is often called a batch *rectifier*.

The basic difference between batch distillation (rectifier) and continuous distilla-

---

[1]This chapter is based on the book by [7].

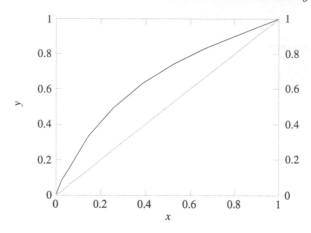

FIGURE 4.1

Equilibrium curve for the CS_2 and CCl_4 mixture at 1 atmosphere pressure

tion is that in continuous distillation (Figure 4.2b) the feed is continuously entering
the column, while in batch distillation the feed is charged into the reboiler at the
beginning of the operation. This feed is heated to form vapor. This vapor is passed
through packing or plates where it meets the liquid coming from top (refluxed back)
from the condenser at the top. Mass transfer takes place in contact and one gets a
distillate which is richer in the more volatile component. The reboiler in batch distillation gets depleted over time. The process has an unsteady-state nature. Reflux
ratio, which is a ratio of liquid reflux to distillate is an important operating parameter on which operating condition depends. If reflux ratio is kept constant throughout
the operation this results in variable product composition. On the other hand, in
order to keep the key composition constant, reflux ratio needs to be varied. There
is a third policy of operation known as the optimal reflux policy that is neither the
constant reflux policy nor the variable reflux policy. Instead, this operating policy
exploits the difference between the two operating modes. Thus, the optimal reflux
policy is essentially a *trade-off* between the two operating modes, and is based on
the ability of the process to yield the most profitable operation. Thus, there are
three operating conditions for batch distillation.

- Constant reflux and variable product composition,

- Variable reflux and constant product composition of the key component, and

- Optimal reflux and optimal product composition.

    The flexible and transient nature of batch distillation allows to configure the
column in a number of different ways, some of which are shown in Figure 4.3 [36].

    The column in Figure 4.3a, as explained, is a conventional batch distillation
column with the reboiler at the bottom and the condenser at the top. A single column
can be used to separate several products using the multi-fraction operation of batch
distillation presented in Figure 4.3b. Some cuts may be desired and others may be
intermediate products. These intermediate fractions can be recycled to maximize
profits and/or minimize waste generation. Figure 4.3c shows a periodic operation

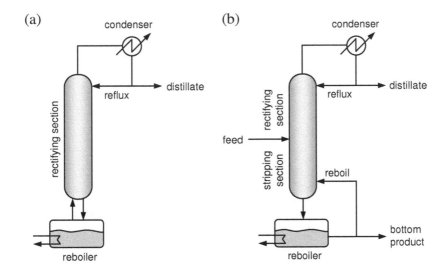

FIGURE 4.2
Types of distillation processes: (a) batch distillation and (b) continuous distillation.

in which each charge consists of a fresh feed stock mixed with the recycled off-specification material from the previous charge. Figure 4.3d represents a stripping column for separating a heavy component as the bottom product where the liquid feed is initially charged into the top. In 1994, Davidyan et al. [37] presented a batch distillation column that has both stripping and rectifying sections embedded in it (Figure 4.3e). This column is called the middle vessel column. Although this column has not been investigated completely, recent studies demonstrated that it provides added flexibility for the batch distillation operation. Recently [38] described a new column configuration called a multivessel column (Figure 4.3f) and showed that the column can obtain purer products at the end of a total reflux operation. These emerging column designs play an important role in separation of complex systems such as azeotropic, extractive, and reactive batch distillation systems. Although all these column configurations are important, we will focus on the conventional batch column in this chapter.

4.1 Early Theoretical Analysis

This section presents early theoretical analysis of simple distillation, which was first analyzed by [35].

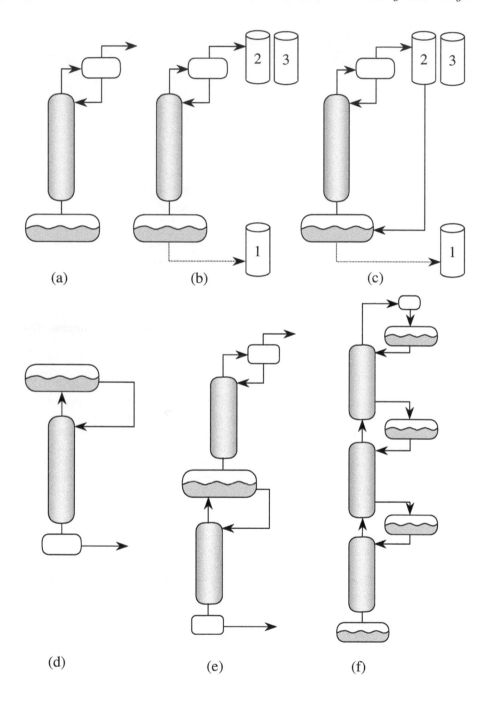

FIGURE 4.3
Various configurations of batch distillation column.

4.1.1   Simple Distillation

The analysis of simple distillation presented by Rayleigh in 1902 marks the earliest theoretical work on batch distillation. Simple distillation, also called Rayleigh distillation or differential distillation, is the most elementary example of batch distillation. In this distillation system, the vapor is removed from the still during each time interval and is condensed in the condenser. The vapor is richer in the more volatile component than the liquid remaining in the still. Over time, the liquid remaining in the still begins to experience a decline in the concentration of the more volatile component, while the distillate collected in the condenser becomes enriched in the more volatile component. No reflux is returned to the still, and no plates or packing materials are provided inside the column. Therefore, various operating policies are not applicable to this distillation system.

The early analysis of this process for a binary system, proposed by Rayleigh, is given below. Let F be the initial binary feed to the still (moles), and x_F be the mole fraction of the more volatile component A in the feed. Let B be the amount of compounds remaining in the still, x_B the mole fraction of component A in the still, and x_D the mole fraction of component A in the vapor phase. The differential material balance for component A can then be written as:

$$x_D \, dB = d(B \, x_B) = B \, dx_B + x_B \, dB, \qquad (4.1)$$

giving:

$$\int_F^B \frac{dB}{B} = \int_{x_F}^{x_B} \frac{dx_B}{x_D - x_B}, \qquad (4.2)$$

or:

$$\ln\left(\frac{B}{F}\right) = \int_{x_F}^{x_B} \frac{dx_B}{x_D - x_B}. \qquad (4.3)$$

In this simple distillation process, it is assumed that the vapor formed within a short period is in thermodynamic equilibrium with the liquid. Hence, the vapor composition x_D is related to the liquid composition x_B by an equilibrium relation of the form $x_D = f(x_B)$. The exact relationship for a particular mixture may be obtained from a thermodynamic analysis depending on temperature and pressure. For a system following the ideal behavior given by Raoult's law, the equilibrium relationship between the vapor composition y (or x_D) and liquid composition x (or x_B) of the more volatile component in a binary mixture can be approximated using the concept of constant relative volatility (α), and is given by:

$$y = \frac{\alpha x}{(\alpha - 1)x + 1} \qquad (4.4)$$

Substitution of the above equation in Equation 4.3 results in:

$$\ln\left(\frac{B}{F}\right) = \frac{1}{\alpha - 1} \ln\left[\frac{x_B(1 - x_F)}{x_F(1 - x_B)}\right] + \ln\left[\frac{1 - x_F}{1 - x_B}\right] \qquad (4.5)$$

Although the analysis of simple distillation historically represents the theoretical start of batch distillation research, a complete separation using this process is impossible unless the relative volatility of the mixture is infinite. Therefore, the application of simple distillation is restricted to laboratory scale distillation, where high purities are not required, or when the mixture is easily separable.

Example 4.1: A mixture of components A and B with 0.4 mole fraction of A and relative volatility of 2.0 is distilled in a simple distillation. The feed is 100 moles, and 40% of mixture is distilled. Find the distillate composition at the end of operation.

Solution: Since 40% of feed is distilled, the residue amount in the reboiler is 60 moles. The bottom composition can be found using Equation 4.5 as given below.

$$\ln\left(\frac{60}{100}\right) = \frac{1}{2-1}\ln\left[\frac{x_B(1-0.4)}{0.4(1-x_B)}\right] + \ln\left[\frac{1-0.4}{1-x_B}\right] \Rightarrow x_B = 0.3138$$

Then the distillate composition can be obtained using Equation 4.4, resulting in distillate composition of 0.4780. Initial distillate composition corresponding to $x_F = 0.4$ (again from the same Equation 4.4) is 0.5700. Therefore, the average distillate composition is between 0.5700 and 0.4780. As can be seen that this distillate composition is very low for a separation process, simple distillation cannot be used for separation in real practice. To obtain products with high purity, multi-stage batch distillation is used.

4.1.2   Operating Modes

As stated earlier, the two basic modes of batch distillation are (1)constant reflux, and (2)variable reflux, resulting in variable distillation composition and constant distillate composition of the key component, respectively. The third operating mode, optimal reflux or optimal control is the trade-off between the two operating modes.

It is easy to understand the two basic modes using binary distillation. This involves use of McCabe-Thiele graphical method described below.

McCabe-Thiele Graphical Method

Although simple distillation marks the first analysis of batch distillation process, the graphical analysis presented by [39] provided the basis for analyzing batch distillation operating modes. The difference between simple distillation and batch distillation operations is the relation between the distillate composition x_D and the bottom composition x_B due to the presence of reflux and column internals as shown in Figure 4.4. If one assumes constant molal overflow ($L_j = L_{j-1} = .. = L_0 = L$ and $V_j = V_{j-1} = .. = V_1 = V$), then McCabe-Theile's method can be used.

In the McCabe-Thiele method, the overall material balance (with no liquid holdup) on the plate is considered from the condenser to the j-th plate. This leads to the following operating line equation (Equation 4.6). Note that in this procedure, we use the theoretical number of stages or plates. If you divide the number of theoretical plates with the plate efficiency, we get the actual plates in a plate column, and if you multiply these theoretical stages with the Height Equivalent to a Theoretical Plate (HETP) then you get the actual height of a packed column.

$$y_j = \frac{R}{R+1}x_{j-1} + \frac{1}{R+1}x_D \tag{4.6}$$

where $R = L/\frac{dD}{dt}$ and $V = (R+1)\frac{dD}{dt}$.

This operating equation represents a line through the point y_j ($x_{j-1} = x_D$)

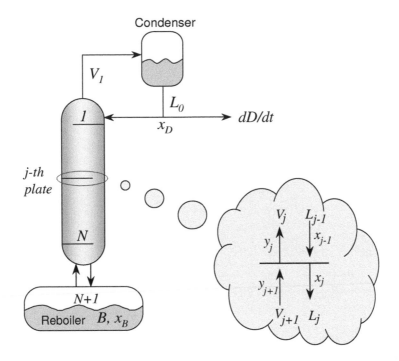

FIGURE 4.4
Schematic of a batch distillation column.

with a slope of $R/(R+1)$. From $(x_0 = x_D, y_1 = x_D)$ draw a horizontal line to equilibrium curve to find the x_1, for this x_1, $y_j + 1$ can be found out from the operating line (vertical line, Equation 4.6). So the procedure is, starting from point (x_D, x_D), Equation 4.6 and the equilibrium curve (e.g., Equation 4.4) between y_j and x_j can be recursively used from the top plate 1 to the reboiler (the reboiler can be considered as the $(N+1)$-th plate) as shown in Figure 4.5 (where $N + 1$ is equal to 6). This procedure relates the distillate composition x_D to the still composition x_B through the number of stages, N.

Constant Reflux Mode

Smoker and Rooc[40] presented the first analysis of the constant reflux operation of a binary batch distillation with no holdup. They used the Rayleigh equation in conjunction with the McCabe-Thiele graphical method to capture the dynamics of the batch distillation column. In their procedure, the relationship between x_D and x_B is recursively determined by the McCabe-Thiele graphical method. Then, the right hand side of the Rayleigh equation (Equation 4.3) is integrated graphically by plotting $1/(x_D - x_B)$ versus x_B. The area under the curve between the feed composition x_F and the still composition x_B now gives the value of the integral, which is $\ln(B/F)$. The average composition of the distillate can be obtained from

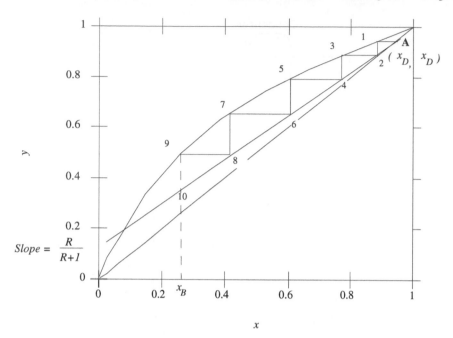

FIGURE 4.5
McCabe-Thiele method

the following equation:

$$x_{D,avg} = \frac{F\,x_F - B\,x_B}{F - B} \tag{4.7}$$

Although Smoker and Rose presented the calculation method independent of time, the time can be introduced through the vapor boilup rate V of the reboiler. The resulting equation for determining total batch time is given by:

$$dB = \frac{V}{R+1}dt \;\Rightarrow\; T = \frac{R+1}{V}(F - B) = \frac{R+1}{V}D \tag{4.8}$$

This operation policy is the easiest one and is commonly used. The following example illustrates this policy.

Example 4.2: Consider the same mixture presented in Example 4.1 but use batch distillation column with five theoretical plates and reflux ratio equal to 5.630.

a) Find the distillate and still composition when 40 percent of the mixture is distilled similar to Example 4.1.

b) If the initial feed is 100 moles and vapor boilup rate is 143.1 moles/hr, what is the total time required to complete the distillation operation?

Solution: The distillation operation requires 40 percent of the mixture to be distilled, which corresponds to 60 percent of the original feed F remaining in the reboiler, i.e., $B = 0.6F$. The problem is that of determining the distillate and still composition, for which the Rayleigh equation (Equation 4.3) applies. To attain the specified requirements, the left hand side of the Rayleigh equation can be calculated as:

$$\ln\left(\frac{F}{B}\right) = \ln\left(\frac{1}{0.6}\right) = 0.5108$$

For various values of x_D the operating lines are drawn and the bottom composition is calculated using the procedure shown in Figure 4.5. From trial and error, we discover that the initial value of x_D corresponding to the still composition of $x_F = 0.4$ is 0.9471. The rest of the x_D are chosen below this composition, and various values of x_B, $x_D - x_B$, $\frac{1}{x_D - x_B}$ are obtained.

Values of x_B versus $\frac{1}{x_D - x_B}$ are plotted in Figure 4.6 from which $\int_{x_F}^{x_B} \frac{dx_B}{x_D - x_B}$ is obtained for each value of x_B. The operation is stopped when the integral is equal to 0.5108.

a) The distillate composition x_D at the end of the operation is 0.5681, and the still composition x_B is 0.1000. The average distillate composition is obtained using the following material balance equations.

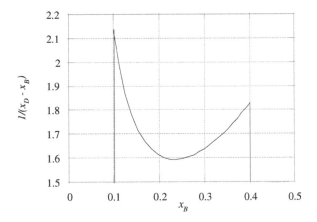

FIGURE 4.6
Graphical integration for Example 4.2

$$\int_0^D x_D dD = \int_F^B x_B dB$$

$$x_{Dav}\int_0^D dD = Fx_F - Bx_B$$

$$x_{Dav} = \frac{Fx_F - Bx_B}{D}$$

$$x_{Dav} = \frac{100 \times 0.4 - 60 \times 0.1}{40} = 0.8500$$

The average distillate composition is 0.85 for the key component which is much larger than obtained in simple distillation.

b) The time required for the distillation is given below.

$$T = \frac{R+1}{V} \times D = 1.8532 \text{ hrs}$$

Variable Reflux

In 1937, Bogart [41] presented the first analysis of the variable reflux policy for a binary system. The steps involved in the calculation procedure for the variable reflux mode are similar to those in the case of the constant reflux mode; however, in the variable reflux case, the reflux ratio is varied instead of the distillate composition at each step. Moreover, the Rayleigh equation, though valid for the variable reflux condition, takes a simplified form. Since the distillate composition remains constant (remember that we are considering binary systems here) throughout the operation, the Rayleigh equation reduces to the following equation.

$$\frac{B}{F} = \frac{x_D - x_F}{x_D - x_B} \tag{4.9}$$

The second step is to establish the relation between R and x_B using the McCabe-Thiele graphical method. Several values of R are selected, operating lines are drawn through the fixed point (x_D, x_D) with the slope of $R/(R+1)$, and steps are drawn between the operating line and the equilibrium curve to get the bottom composition (x_B). This recursive scheme is repeated until the desired stopping criteria is met, and thus B and x_B can be found at each value of the reflux ratio. The time required for this operation at a given product purity is calculated by plotting the quantity $\frac{(R+1)}{V} \times \frac{F(x_D - x_F)}{(x_D - x_F)^2}$ versus x_B in the following equation and then finding the area under the curve.

$$T = \int_{x_B}^{x_F} \frac{R+1}{V} \frac{F(x_D - x_F)}{(x_D - x_B)^2} \, dx_B \tag{4.10}$$

Example 4.3: Rework the problem in Example 4.2 for the variable reflux mode.

Solution: Since the distillate composition is held constant throughout the variable reflux mode of operation, the distillate composition $x_D = x_{Dav} = 0.85$. For the various iterates of R, we obtain the corresponding values of x_B. The value of the amount of product distilled at each x_B is also calculated using the Rayleigh equation for the variable reflux condition (Equation 4.9).

$$D = F(1 - \frac{x_D - x_F}{x_D - x_B})$$

a) The operation is stopped when the amount of product collected is greater than or equal to 40. The still composition at $D = 40$ is found to be equal to 0.10.

b) The time required for this operation is calculated by plotting the quantity $\frac{R+1}{V} \frac{F(x_D - x_F)}{(x_D - x_B)^2}$ versus x_B and then finding the area under the curve between x_B equal to 0.4 (Figure 4.7) and 0.1. The time required is found to be 1.995 hrs.

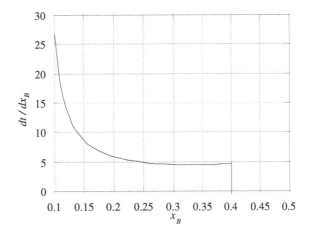

FIGURE 4.7

Graphical integration for calculation of batch time for Example 4.3

Optimal Reflux Policy

The optimal reflux mode is a reflux profile that optimizes the given indices of column performance chosen as the objectives. The indices used in practice generally include the minimum batch time, maximum distillate, or maximum profit functions. This reflux policy is essentially a *trade-off* between the two operating modes, and is based on the ability to yield the most profitable operation from optimal performance. The calculation of this policy is a difficult problem and relies on optimal control theory and is discussed in Chapter 13 on optimization and optimal control. For example, consider the following reflux profile for the same separation given in Examples 4.2 and 4.3.

| x_D | .9019 | .8980 | .8899 | .8749 | .8455 | .8265 | .8010 | .7635 | .7018 |
| R | 5.472 | 5.349 | 5.237 | 5.138 | 4.980 | 5.348 | 5.966 | 6.839 | 7.623 |

Remember we are using the same batch and the same column for this operation. Since in this case neither distillate composition nor reflux ratio is constant, the following procedure is used for integration of the Rayleigh equation.

For each pair of values of x_D and R, operating lines are drawn with a slope equal to $\frac{R}{R+1}$ and passing through the point (x_D, x_D). Plate to plate calculations are used to obtain the corresponding values of x_B.

As stated earlier, the basic batch distillation column satisfies the Rayleigh equation. Therefore we can use the same equation to calculate the total amount of distillate, as was done for the constant reflux condition. Values of x_B, $x_D - x_B$, $1/x_D - x_B$ are obtained for each operating line. Values of x_B versus $1/x_D - x_B$ are plotted from which the right hand side of the Rayleigh equation is obtained as the area under the curve and is found to be 0.5108, which corresponds to $\ln[\frac{B}{F}]$. Hence the total amount of distillate given by $F(1 - \frac{B}{F})$ is equal to 40 moles, which is what we obtained in the variable reflux and constant reflux operations in Examples 4.2 and 4.3.

We resort to the basic mass balance equation to obtain the average distillate composition used in the case of constant reflux mode.

$$\int_0^D x_D\,dD \;=\; \int_F^B x_B\,dB$$

$$x_{Dav}\int_0^D dD \;=\; Fx_F \;-\; Bx_B$$

$$x_{Dav} \;=\; \frac{Fx_F \;-\; Bx_B}{D}$$

$$x_{Dav} \;=\; \frac{100\times0.4 \;-\; 60\times0.1}{40} \;=\; 0.8500$$

The average distillate composition in this case is found to be 0.85, which is again the same as that of the two operating modes.

Similarly, the time requirement for this operation is derived from the basic material balance equations as was done in the case with the variable reflux condition. The time T required when neither distillate composition nor reflux is constant is found to be

$$T \;=\; \int_{x_B}^{x_F} \frac{B}{V}\frac{R+1}{x_D\;-\;x_B}\,dx_B \tag{4.11}$$

The value of T is found to be 1.686 hrs, smaller than for the constant and variable reflux modes of operation.

It can be seen that this reflux profile requires the least amount of time to obtain the same amount of product with the same purity. This policy is neither variable reflux nor constant reflux but is in between the two as shown in Figure 4.8. The policy which was specified in the above example is the optimal policy obtained by Coward in 1967[42].

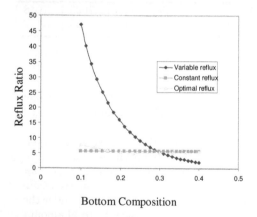 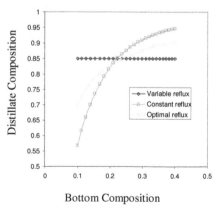

FIGURE 4.8
The three operating modes

4.2   Hierarchy of Models

As seen in earlier sections, the earlier models of the batch rectifier were built on assumptions of negligible liquid holdup and ideal binary systems. Computers have played an important role in relaxing these assumptions, especially the negligible holdup assumption. Distefano analyzed the numerical differential equations for multicomponent batch distillation in 1968 for the first time[43]. The rigorous models of batch distillation in current state of the art computer packages are based on his pioneering work. However, it was also acknowledged that due to the severe transients in batch distillation, a hierarchy of models is necessary to capture the dynamics of this flexible operation[44, 36]. This section presents the hierarchy of models ranging from the rigorous model (similar to the one presented by Distefano) to the simplest shortcut model[45].

4.2.1   Rigorous Model

A rigorous model in batch distillation involves consideration of column dynamics along with the reboiler and condenser dynamics. Distefano presented detailed analysis of the characteristics of differential mass and energy balances associated with the complete dynamics of a multicomponent batch distillation column. Distefano's work forms the basis for almost all of the later work on rigorous modeling of batch distillation columns, and this model is presented below.

For an arbitrary plate j (Figure 4.4), the total mass, component, and energy balances yield the governing equations are summarized in Table 4.1. This table lists all the equations involved in the dynamic analysis of the batch column and the assumptions behind these equations. Note that energy balance equations are difference equations instead of differential equations, this is due to the fact that flowrates are much larger as compared to the enthalpy changes.

The system of equations governing the batch distillation process is difficult to solve as the plate holdup is generally much smaller than reboiler holdup resulting in severe transients. Stiff equation solver is necessary to solve these type of equations. The stiffness of the system is reduced considerably when one considers zero plate holdup. This results in semirigorous model for batch distillation. This model is similar to what was used earlier with McCabe-Theile method (except with additional energy balance equations whenever necessary).

4.2.2   Low Holdup Semirigorous Model

For columns where the plate dynamics are significantly faster than the reboiler dynamics (due to very small plate holdups and/or wide boiling components), the stiff integrator often fails to find a solution. The solution to this problem is to split the system into two levels: (a) the reboiler, where the dynamics are slower, can be represented by differential equations (Equations 4.12-4.13), and (b) the rest of the column can be assumed to be in the *quasi-steady* state. Thus, the composition changes in the condenser and accumulator ($dx_D^{(i)}/dt$), the composition changes on plates ($dx_j^{(i)}/dt$), and the enthalpy changes in the condenser and on plates ($\delta_t I_D$

TABLE 4.1
The complete column dynamics

Assumptions
- Negligible vapor holdup,
- Adiabatic operation,
- Theoretical plates,
- Constant molar holdup,
- Finite difference approximations for the enthalpy changes.

Composition Calculations

Condenser and Accumulator Dynamics

$$\frac{dx_D^{(i)}}{dt} = \frac{V_1}{H_D}(y_1^{(i)} - x_D^{(i)}), i = 1, 2, \ldots, n$$

Plate Dynamics

$$\frac{dx_j^{(i)}}{dt} = \frac{1}{H_j}[V_{j+1}y_{j+1}^{(i)} + L_{j-1}x_{j-1}^{(i)} - V_j y_j^{(i)} - L_j x_j^{(i)}],$$
$$i = 1, 2, \ldots, n; j = 1, 2, \ldots, N$$

Reboiler Dynamics

$$\frac{dx_B^{(i)}}{dt} = \frac{1}{B}[L_N(x_N^{(i)} - x_B^{(i)}) - V_B(y_B^{(i)} - x_B^{(i)})], i = 1, 2, \ldots, n$$

Flowrate Calculations

At the Top of the Column

$$L_0 = R\frac{dD}{dt}; V_1 = (R+1)\frac{dD}{dt}$$

On the Plates

$$L_j = V_{j+1} + L_{j-1} - V_j; j = 1, 2, \ldots, N$$
$$V_{j+1} = \frac{1}{J_{j+1} - I_j}[V_j(J_j - I_j) + L_{j-1}(I_j - I_{j-1}) + H_j \delta I_j]$$
$$j = 1, 2, \ldots, N$$

At the Bottom of the Column

$$\frac{dB}{dt} = L_N - V_B$$

Heat Duty Calculations

Condenser Duty

$$Q_D = V_1(J_1 - I_D) - H_D \delta_t I_D$$

Reboiler Duty

$$Q_B = V_B(J_B - I_B) - L_N(I_N - I_B) + B\delta_t I_B$$

Thermodynamics Models

Equilibrium Relations

$$y_j^{(i)} = f((x_j^{(k)}, k = 1, \ldots, n), TE_j, P_j)$$

Enthalpy Calculations

$$I_j = f((x_j^{(k)}, j = 1, \ldots, n), TE_j, P_j)$$
$$J_j = f((y_j^{(k)}, j = 1, \ldots, n), TE_j, P_j)$$

and $\delta_t I_j$) in Table 4.1 can be assumed to be zero. These equations can be integrated using Runge-Kutta method. This results in a zero holdup model, so this approach can be used for simulating the semirigorous model of batch distillation.

$$\frac{dB}{dt} = -\frac{V}{R+1}, \qquad B_0 = F \qquad (4.12)$$

$$\frac{dx_B^{(k)}}{dt} = \frac{V}{(R+1)B}(x_B^{(k)} - x_D^{(k)}), \qquad x_{B0}^{(k)} = x_F^{(k)} \qquad (4.13)$$

The holdup effects can be neglected in a number of cases where this model approximates the column behavior accurately. This model provides a close approximation to the Rayleigh equation, and for complex systems (e.g., azeotropic systems) the synthesis procedures can be easily derived based on the simple distillation residue curve maps (trajectories of composition). However, note that this model involves an iterative solution of nonlinear plate-to-plate algebraic equations, which can be computationally less efficient than the rigorous model.

4.3   Shortcut Model

The rigorous model of batch distillation operation involves a solution of several stiff differential equations and the semirigorous model involves a set of highly nonlinear equations. The computational intensity and memory requirement of the problem increase with an increase in the number of plates and components. The computational complexity associated with these models does not allow us to derive global properties such as feasible regions of operation, which are critical for optimization, optimal control, and synthesis problems. Even if such information is available, the computational costs of optimization, optimal control, or synthesis using these models are prohibitive. One way to deal with these problems associated with these models is to develop simplified models such as the shortcut model.

The shortcut model of batch distillation proposed by [45] is based on the assumption that the batch distillation column can be considered equivalent to a continuous distillation column with changing feed at any instant. Since continuous distillation theory is well-developed and tested, the shortcut method of continuous distillation is modified for batch distillation, and the compositions are updated using a finite-difference approximation for the material balance (based on the Rayleigh equation). The other assumptions of the shortcut method include constant molar overflow and negligible plate holdups. As described earlier, the functional relationship between the distillate composition x_D and the bottom composition x_B is crucial for the simulation, and the FUG (Fenske-Underwood-Gilliland) method is used for estimating this relation. These equations shown in Table 4.2 along with the material balance equations (Equations 4.12-4.13) are used to simulate the three operating conditions of batch distillation.

At any instant of time, Equations 4.12, 4.13, and the differential material balance equation can be used to calculate the bottom composition of all the components. The following procedure is then used to calculate the distillate composition at that instant. The procedure is repeated at each time step until the stopping criterion is

TABLE 4.2
Time implicit model equations for the shortcut method

Variable Reflux	Constant Reflux	Optimal Reflux
Differential Material Balance Equation		
$x_{B_{new}}^{(i)} = x_{B_{old}}^{(i)} + \dfrac{\Delta x_B^{(k)}\left(x_D^{(i)} - x_B^{(i)}\right)_{old}}{\left(x_D^{(k)} - x_B^{(k)}\right)_{old}}, i = 1, 2, \ldots, n$		
Hengstebeck-Geddes Equation		
$x_D^{(i)} = \left(\dfrac{\alpha_i}{\alpha_k}\right)^{C_1} \dfrac{x_D^{(k)}}{x_B^{(k)}} x_B^{(i)}, i = 2, 3, \ldots, n$		
Unknowns		
R, C_1	$C_1, x_D^{(k)}$	$R, C_1, x_D^{(k)}$
Summation of Fractions		
$\sum_{i=1}^{n} x_D^{(i)} = 1$		
C_1 *Estimation*	$x_D^{(k)}$ *estimation*	
$\sum_{i=1}^{n} \left(\dfrac{\alpha_i}{\alpha_k}\right)^{C_1} \dfrac{x_D^{(k)}}{x_B^{(k)}} x_B^{(i)} = 1$	$x_D^{(k)} = \dfrac{1}{\sum_{i=1}^{n} \left(\dfrac{\alpha_i}{\alpha_k}\right)^{C_1} \dfrac{x_B^{(i)}}{x_B^{(k)}}}$	
Fenske Equation		
$N_{min} \approx C_1$		
Underwood Equations		
$\sum_{i=1}^{n} \dfrac{\alpha_i x_B^{(i)}}{\alpha_i - \phi} = 0$; $R_{min}\, u + 1 = \sum_{i=1}^{n} \dfrac{\alpha_i x_D^{(i)}}{\alpha_i - \phi}$		
Gilliland Correlation		
R *Estimation*		C_1 *Estimation*
$Y = 1 - \exp\left[\dfrac{(1 + 54.4X)(X - 1)}{(11 + 117.2X)\sqrt{X}}\right]$		
$X = \dfrac{R - R_{min}}{R + 1}$; $Y = \dfrac{N - N_{min}}{N + 1}$		
		R *Estimation*
		$R = F(Minimum\ H)$

satisfied. The complete simulation of the shortcut method for the different operating modes is illustrated in the flowchart shown in Figure 4.9.

The constant C_1 in the Hengstebeck-Geddes equation is equivalent to the minimum number of plates, N_{min}, in the Fenske equation. At this stage, the variable reflux operating mode has C_1 and R, the constant reflux has $x_D^{(k)}$ and C_1, and the optimal reflux has $x_D^{(k)}$, C_1, and R as unknowns. Summation of distillate compositions can be used to obtain C_1 for variable reflux and $x_D^{(k)}$ for both constant reflux and optimal reflux operation, and the FUG equations to obtain R for variable reflux and C_1 for both constant reflux and optimal reflux operations. The optimal reflux mode of operation has an additional unknown, R, which is calculated using the concept of optimizing the Hamiltonian, formulated using the different optimal control methods.

The shortcut model is extremely efficient and reasonably accurate for nearly ideal systems and for column with negligible holdup. For columns with severe holdup

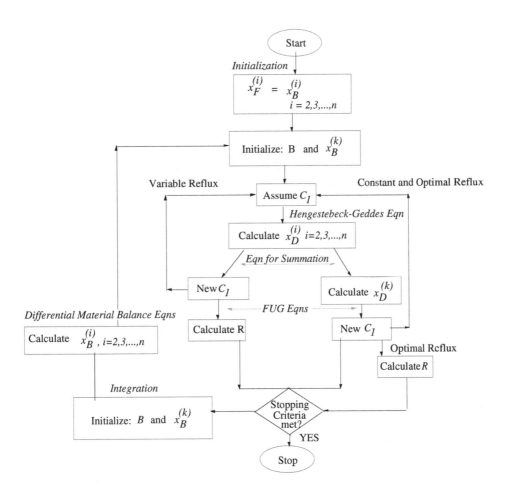

FIGURE 4.9
The flowchart for the shortcut method

effect and highly non-ideal systems, the order of rigorous model can be reduced using collocation approach.

4.4   Optimization and Optimal Control

Literature on the optimization of the batch column is focused mostly on the solution of optimal control problems, which includes optimizing the indices of performance such as maximum distillate, minimum time, and maximum profit. However, literature on optimal design of batch distillation for performing specified operations by using the constant reflux or variable reflux policies is very limited[46].

As stated in Chapter 1, control refers to a closed-loop system where the desired operating point is compared to an actual operating point and a knowledge of the error is fed back to the system to drive the actual operating point towards the desired one. However, the optimal control problems we consider here do not fall under this definition of control. Because the decision variables that will result in optimal performance are time-dependent, the control problems described here are referred to as optimal control problems. Thus, use of the control function here provides an open-loop control. The dynamic nature of these decision variables makes these problems much more difficult to solve as compared to normal optimization where the decision variables are scalar.

The indices of performance used in batch distillation optimal control problems are discussed below.

- **Maximum Distillate Problem** – where the amount of distillate of a specified concentration for a specified time is maximized [47, 48, 49, 50, 51]. This problem can be mathematically represented as follows:

$$\max_{R_t} \quad J = \int_0^T \frac{dD}{dt}\, dt = \int_0^T \frac{V}{R_t + 1}\, dt, \qquad (4.14)$$

subject to the material and energy balances.

[47] were the first to report the maximum distillate problem for binary batch distillation, which was solved using Pontryagin's maximum principle, the dynamic programming method, and the calculus of variations. These methods are described in Chapter 14. [48] extended this optimization model to multicomponent systems and used the shortcut batch distillation model along with the maximum principle for the calculation of the optimal reflux policy. [49] used the orthogonal collocation approach on finite elements and nonlinear programming (NLP) optimization techniques over the shortcut model. Further, they extended this method to the rigorous batch distillation model [51] in which they considered the effect of column holdups on optimal control policy.

- **Minimum Time Problem** – where the batch time needed to produce a prescribed amount of distillate of a specified concentration is minimized [42, 52]. Although there are several different formulations for the minimum time problem, [52] derived the following formulations to establish a unified theory for all the optimal control problems:

$$\min_{R_t} \quad J = \int_0^T \frac{dt^*}{dt} dt. \tag{4.15}$$

where t^* is a dummy variable as a state variable.

- **Maximum Profit Problem** – where a profit function for a specified concentration of distillate is maximized [53, 49, 54, 55].

Much of the recent research on optimal control problems can be classified into this problem. [53] were the first to use the profit function for maximization in batch distillation , and they solved the optimal control problem. The following simple objective function is given by [53]:

$$\max_{R_t,T} \quad J = \frac{DP_r - FC_F}{T + t_s}, \tag{4.16}$$

subject to purity constraints and column modeling equations.

[48] used a different objective function to solve the profit maximization problem under the constant and variable reflux conditions. [49] formulated a new profit function and solved the differential algebraic optimization problem for optimal design and operation. [54] developed a detailed dynamic multifraction batch distillation model and discretized the model using the orthogonal collocation method on finite elements, and finally solved the maximum profit model using a NLP optimizer. [55] considered a rigorous reactive distillation system for the maximum conversion problem, which can also be classified as the maximum profit problem. The detailed dynamic system is then reduced by using polynomial curve fitting techniques and solved by using a NLP optimizer.

A variant of this objective function is to minimize the mean rate of energy consumption when the market size for the product is fixed by the current demand. The objective function is given by [56]:

$$\min \quad J = \frac{\int_0^T Q_R(t)dt}{T + t_s}, \tag{4.17}$$

$$\text{s.t.} \quad x_{D,\text{avg}} \geq x^*,$$

$$D \geq D^*,$$

where Q_R is the reboiler heat duty. They used this objective function for optimal control of multivessel columns for the first time. [57] also presented the optimal operation policy based on energy consumption for the multivessel column.

4.5 Complex Systems

Thermodynamically and kinetically complex systems like azeotropic, extractive, and reactive systems pose additional bottlenecks in design and operation of batch columns. The following sections describe the methods for analyzing these complex systems. These methods also provide heuristics for synthesis of these columns especially in terms of the different cuts obtained in a single column or performance comparison of the complex columns.

4.5.1 Azeotropic Distillation

In a normal distillation column, the vapor becomes steadily richer in the more volatile component as it passes through successive plates. In azeotropic mixtures this steady increase in concentration does not take place, owing to the so-called azeotropic points. For instance, when a mixture of ethyl alcohol and water is distilled, the concentration of the alcohol steadily increases until it reaches 92.4 percent by moles, when the composition of the vapor equals that of the liquid, and no further enrichment occurs (see Figure 4.10). Such a mixture is called an azeotrope and cannot be separated by straightforward distillation.

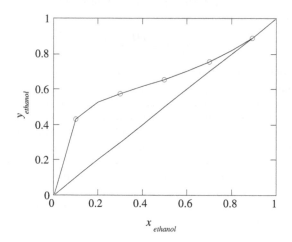

FIGURE 4.10

Vapor-liquid equilibrium curve for the ethanol-water system

Azeotropic distillation is an important and widely used separation technique as a large number of azeotropic mixtures are of great industrial importance. Theoretical studies on azeotropic distillation have mainly centered around methods for predicting the vapor-liquid equilibrium data from liquid solution models and their application to distillation design. However, only during the past two decades has there been a concerted effort to understand the nature of the composition region boundaries. In binary distillation, the barrier is represented by the azeotropic points and in case of a ternary system, they are represented by separatrices. Based on this information, qualitatively similar ideal systems can be represented for these non-ideal azeotropic systems[58, 59]. Doherty & coworkers [60, 61] in their pioneering works proposed several new concepts in azeotropic distillation. They established the use of ternary diagrams and residue curve maps in the design and synthesis of azeotropic continuous distillation columns. In batch distillation, they outlined a synthesis procedure based on the residue curve maps.

The residue curve map graphs the liquid composition paths that are solutions to the following set of ordinary differential equations:

$$\frac{dx_i}{d\xi} = x_i - y_i \qquad i = 1, 2, \ldots, n - 1, \tag{4.18}$$

where n is the number of components in the system, and the independent variable,

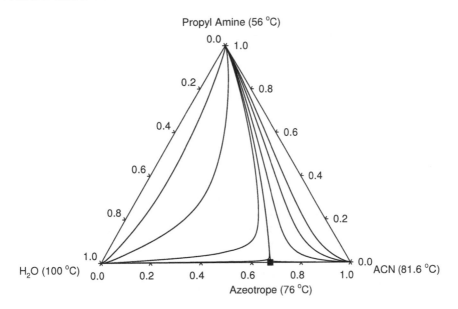

FIGURE 4.11
A residue curve map of the propyl amine-acetonitrile-water system.

warped time (ξ), is a monotonically increasing quantity related to real time. One can see that Equation 4.18 is one form of the Rayleigh equation described earlier. The residue curve map occupies a significant place in the conceptual design stage of column sequencing in continuous distillation, and fractions (cuts) sequencing in batch distillation [60, 61, 62].

Despite the advances in the thermodynamics for predicting azeotropic mixture, feasible distillation boundaries, and sequence of cuts, the azeotropic batch distillation system is still incipient in terms of design, optimization, and optimal control.

Example 4.4: A residue curve map of the propyl amine-acetonitrile (ACN)-water system are given in Figure 4.11. Find the batch distillation regions and define the product cuts for each region.

Solution: Since the curve from propyl amine to the ACN-water azeotrope is distillation barrier, there are two distillation regions in this system. For the left distillation region, the product sequence is propyl amine, ACN-water azeotrope, and water. For the right region, the sequence is propyl amine, ACN-water azeotrope, and ACN. This example shows that conventional distillation cannot obtain pure water and pure ACN cuts at the same time.

4.5.2   Reactive Distillation

Although reactive distillation was acknowledged as a unit operation as early as in the 1920s, it has gained its research interest as an excellent alternative to both reaction and separation since the 1980s. For example, most of the new commercial processes of MTBE (methyl-tertiary-butyl ether, an anti-knocking agent) are based on continuous reactive distillation technologies.

The analysis of a reactive batch distillation model in a staged column was first published by [63]. Using a stiff integrator for the differential and algebraic equations, they presented a numerical solution technique for the esterification of 1-propanol and acetic acid. They argued that in reactive distillation differential energy balances should be included in the solution. [64] developed a new solution technique based on the orthogonal collocation method on the finite element method for the reactive batch distillation of a packed column. The differential contactor model of a packed column, originally designed by [65], was then reduced to the low-order polynomials with desired accuracy. They compared the results with those from the finite difference method and global collocation method for non-reactive packed-bed batch distillation systems, and showed that their approach was more efficient. [66] extended their previous work to the optimal campaign structure for reactive batch distillation, which can offer reasonably sharp separations between successive cuts and reduce the amount of waste off-cuts. To obtain the optimal reflux policies or profiles for the maximum distillate or minimum time problem, multiperiod reflux optimization [67] can be applied. They showed that for the same production rate, the waste generation can be significantly reduced under the optimal campaign structure.

An efficient optimization approach for reactive batch distillation using polynomial curve fitting techniques was presented by [55]. After finding the optimal solution of the maximum conversion problem, polynomial curve fitting techniques were applied over these solutions, resulting in a nonlinear algebraic maximum profit problem that can be efficiently solved by a standard NLP technique. Four parameters in the profit function, which are maximum conversion, optimum distillate, optimum reflux ratio, and total reboiler heat load, were then represented by polynomials in terms of batch time. This algebraic representation of the optimal solution can be used for online optimization of batch distillation.

A dynamic rate-based model for packed-bed batch distillation was recently proposed [68], in which a solid catalyst was used first in the reactive batch distillation modeling. The pilot-scale experiments were conducted with strong anion-exchange resins. The results were compared with the experimental data and with the results from its counterpart, the equilibrium-based model. The rate-based model provides more accuracy, much higher physical significance, and more predicability of the experimental data even though the formulation of the rate-based model is complicated.

4.6   Computer Aided Design Software

It is difficult to analyze batch distillation without using computers due to the two reasons stated before: (a) the process is time varying, and one has to resort to complex numerical integration techniques and different simulation models for obtaining the transients, and (b) this ever-changing process also provides flexibility in operating and configuring the column in numerous ways. Based on the current state of the art in batch distillation techniques and computer simulation technology, Table 4.3 identifies the required functionality and the rationale behind it.

There are several commercial software packages for simulations, optimizations, and/or optimal controls of batch distillation (see Table 4.3). These include Bdist-SimOPT (Batch Process Technologies), BatchSim(Simulation Sciences), BatchFrac (Aspen Technology, based on [69]), and MultiBatchDS (Batch Process Research Company). Bdist-SimOPT and MultiBatchDS are derived from the academic package BATCHDIST [45]. Most of these packages except MultiBatchDS are usually limited to conventional systems as they were developed in early or late, 80s.

4.7 Summary

This chapter presented analysis of batch distillation beginning from the first theoretical analysis in 1902. There are three operating modes of conventional batch distillation column, namely, constant reflux and variable product composition, variable reflux and constant product composition of the key component, and optimal reflux policy. There is a hierarchy of models available for batch distillation. These include shortcut model for feasibility, design, optimization, and synthesis, semirigorous model with negligible holdup, rigorous model including all transients with holdup and reduced model based on collocation method. There are numbers of software packages available for batch distillation design, simulation, and optimization.

Notations

B	amount of bottom residue [mol]
$\frac{dB}{dt}$	bottom product flow rate or change of bottom product [mol/time]
C_1	constant in the Hengstebeck–Geddes equation for conventional batch column
D	amount of distillate [mol]
$\frac{dD}{dt}$	distillate rate [mol/time]
E	entrainer feed rate [mol/time]
F	amount of feed [mol]
H_j	molar holdup on plate j [mol]
H_0, H_D	condenser holdup [mol]
I_D	enthalpy of the liquid in the condenser [mass length^2 time^{-2}/mol]
I_j	enthalpy of the liquid stream leaving plate j [mass length^2 time^{-2}/mol]
J_j	enthalpy of the vapor stream leaving plate j [mass length^2 time^{-2}/mol]
L_j	liquid stream leaving plate j [mol/time]
L_0	liquid reflux at the top of the column [mol/time]
n	number of components
N	number of plates
$N_{min}]$	minimum number of plates

q'	ratio of the top vapor flow rate to the bottom vapor flow rate
Q_R	reboiler heat duty [mass length^2 time^{-2}]
R	reflux ratio ($= L/D$)
$R_{min\ g}$	minimum reflux ratio given by the Gilliland correlation
$R_{min\ u}$	minimum reflux ratio given by the Underwood equations
R_t	reflux ratio as a function of time
T	batch time [time]
V_j	vapor stream leaving plate j [mol/time]
x	liquid-phase mole fraction
x_B	mole fraction of liquid in the reboiler
x_D	mole fraction of the distillate
$x_{D,avg}$	average distillate mole fraction
x_F	liquid-phase mole fraction of the feed
y	vapor-phase mole fraction

Greek Letters:

α	relative volatility

TABLE 4.3
Comparison of software packages

| Features | CHEMCAD BATCH | BATCHSEP | MultiBatchDS | CRANIUM |
	CHEMCAD	ASPEN PLUS		
Databank				
Operations				
Constant Reflux	Yes	Yes	Yes	Yes
Variable Reflux	No	Yes	Yes	Yes
Optimal Reflux	No	No	Yes	Yes
Optimal Reflux-Fixed Equation	No	Yes	Yes	Yes
Models				
Shortcut	No	No	Yes	Yes
Low Holdup Rigorous/Semirigorous	Yes	No	Yes	Yes
Reduced Order	No	No	Yes	Yes
Rigorous	Yes	Yes	Yes	Yes
Configurations				
Rectifier	Yes	Yes	Yes	Yes
Semi-batch	No	Yes	Yes	Yes
Recycle waste cut	No	Yes	Yes	Yes
Stripper	No	No	Yes	Yes
Middle Vessel Column	No	No	Yes	Yes
Options				
Design Feasibility	No	No	Yes	Yes
Optimization	No	Yes	Yes	Yes
Reactive Distillation	No	Yes	Yes	Yes
3 phase Distillation	Yes	Yes	Yes	Yes
Uncertainty Analysis	No	No	Yes	Yes

5

Optimization and Optimal Control

CONTENTS

Numerical optimization[1] plays an important role in batch processing. Whether to find maximum yield in the reactor, or maximum distillation in batch distillation, or optimal schedule for batch processing, optimization and optimal control methods are extensively used. In general, the problems in batch and bio processing are large scale problems where analytical solutions are difficult. Hence numerical optimization methods are necessary.

A general optimization problem can be stated as follows.

$$\text{Optimize} \quad Z = z(x) \qquad (5.1)$$
$$x$$

subject to

$$h(x) = 0 \qquad (5.2)$$

$$g(x) \leq 0 \qquad (5.3)$$

The goal of an optimization problem is to determine the decision variables x that optimize the objective function Z, while ensuring that the model operates within established limits enforced by the set of equality constraints h (Equation 5.2) and inequality constraints g (Equation 5.3).

Figure 5.1 illustrates schematically the iterative procedure employed in a numerical optimization technique. As seen in the figure, the optimizer invokes the model with a set of values of decision variables x. The model simulates the phenomena and calculates the objective function and constraints. This information is utilized by the optimizer to calculate a new set of decision variables. This iterative sequence is continued until the optimization criteria pertaining to the optimization algorithm are satisfied.

---

[1]This chapter is based on the optimization book by [70].

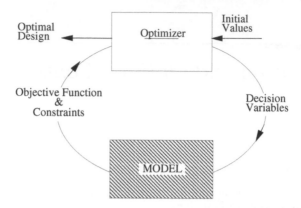

FIGURE 5.1
Pictorial representation of the numerical optimization framework

5.1 Optimization Problems and Software

Optimization problems can be divided into the following broad categories depending on the type of decision variables, objective function(s), and constraints:

- Linear programming (LP): The objective function and constraints are linear. The decision variables involved are scalar and continuous. Examples of LP problems involved in batch processing are scheduling and planning, and blending problems.

- Nonlinear programming (NLP): The objective function and/or constraints are nonlinear. The decision variables are scalar and continuous. Equipment design and operation problems are NLP problems.

- Integer programming (IP): The decision variables are scalars and integers.

- Mixed integer linear programming (MILP): The objective function and constraints are linear. The decision variables are scalar; some of them are integers while others are continuous variables. Again, scheduling and planning problems can be categorized in this category.

- Mixed integer nonlinear programming (MINLP): A nonlinear programming problem involving integer as well as continuous decision variables. Process synthesis problems where simultaneously configuration of the plant and equipment designs are involved are MINLP problems.

- Discrete optimization: Problems involving discrete (integer) decision variables. This includes IP, MILP, and MINLPs.

- Optimal control: The decision variables are vectors. Because of unsteady state nature, these problems are commonly encountered in batch processing.

    There are large numbers of software codes available for numerical optimization. Examples of these include solvers such as MINOS, CPLEX, CONOPT, and NPSOL. Also, many mathematical libraries, such as NAG, OSL, IMSL, and HARWELL have different optimization codes embedded in them. Popular software packages such as

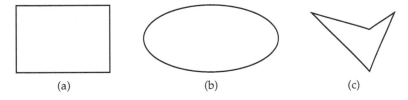

    (a)   (b)   (c)

FIGURE 5.2
Examples of convex and non-convex sets

EXCEL, MATLAB, and SAS also have some optimization capabilities. There are algebraic modeling languages like AMPL, LINGO, AIMMS, GAMS, and ISIGHT specifically designed for solving optimization problems and software products such as Omega and Evolver have spreadsheet interfaces. Furthermore, the Internet provides a great source of information. A group of researchers at Argonne National Laboratory and Northwestern University launched a project known as the Network-Enabled Optimization System (NEOS). Its associated Optimization Technology Center maintains a website at:

$$\text{http://www.mcs.anl.gov/otc/}$$

which includes a library of freely available optimization software, a guide to software selection, educational material, and a server that allows online execution [71]. Also, the site:

$$\text{http://OpsResearch.com/OR-Objects}$$

includes data structures and algorithms for developing optimization applications.

5.2 Convex and Concave Functions

Convexity or concavity of functions is an important property in numerical optimization.

    A set of points S is a convex set if the line segment joining any two points in the space S is wholly contained in S. In Figure 5.2, a and b are convex sets, but c is not a convex set.

    Mathematically, S is a convex set if, for any two vectors x_1 and x_2 in S, the vector $x = \lambda x_1 + (1-\lambda)x_2$ is also in S for any number λ between 0 and 1. Therefore, a function $f(x)$ is said to be strictly convex if, for any two distinct points x_1 and x_2, the following equation applies:

$$f(\lambda x_1 + (1-\lambda)x_2) \; < \; \lambda f(x_1) + (1-\lambda)f(x_2) \tag{5.4}$$

Conversely, a function $f(x)$ is strictly concave if $-f(x)$ is strictly convex.

A global optimum can be easily obtained, if the following conditions apply.

- Maximization: The objective function should be concave and the solution space should be a convex set.

- Minimization: The objective function should be convex and the solution space should be a convex set.

5.3   Linear Programming

Linear programming problems involve linear objective function and linear constraints. The LP optimum lies at a vertex of the feasible region, which is the basis of the simplex method. LP can have 0 (infeasible), 1, or infinite (multiple) solutions. The set of all feasible solutions to a linear programming problem is a convex set. Therefore, a linear programming optimum is a global optimum.

Example 5.1: A batch product manufacturer sells products A and B. The profit from A is \$12/kg and from B \$7/kg. The available raw materials for the products are 100kg of C and 80 kg of D. To produce one kilogram of A, the manufacturer needs 0.5kg of C and 0.5kg of D. To produce one kilogram of B, the manufacturer needs 0.4kg of C and 0.6kg of D. The market for product A is 60kg and for B 120kg. How much raw material should be used to maximize the manufacturer's profit?

Solution: Let x_1 be the amount of A & x_2 be the amount of B
The LP form is:

$$\max \quad Profit\ z \ = \ 12x_1 + 7x_2 \tag{5.5}$$
$$0.5x_1 + 0.4x_2 \ \leq \ 100 \text{Available C} \tag{5.6}$$
$$0.5x_1 + 0.6x_2 \ \leq \ 80 \text{Available D} \tag{5.7}$$
$$x_1 \ \leq \ 60 \tag{5.8}$$
$$x_2 \ \leq \ 120 \tag{5.9}$$
$$x_1, x_2 \geq 0$$

The problem is a two-variable LP problem, which can be easily represented in a graphical form. Figure 5.3 shows constraints (Equation 5.6) through (Equation 5.9), plotted as four lines by considering the four constraints as equality constraints. Therefore, these lines represent the boundaries of the inequality constraints. In the figure, the inequality is represented by the points on the other side of the hatched lines. The objective function lines are represented as dashed lines (isoprofit lines). It can be seen that the optimal solution is at the point $x_1 = 60\ kg$; $x_2 = 83.33\ kg$, a point at the intersection of constraints 5.7, 5.8 and one of the isoprofit lines. The objective function value at the optimum is $z = \$1303$.

Sensitivity Analysis:

The sensitivity of the linear programming solution is expressed in terms of shadow prices (dual price/simplex multipliers) and opportunity (reduced) cost. A shadow

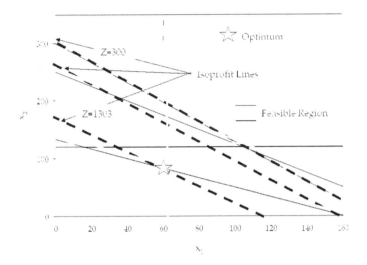

FIGURE 5.3
Linear programming graphical representation, Exercise 5.1

price is the rate of change (increase in the case of maximization and decrease in the case of minimization) of the optimal value of the objective function with respect to a particular constraint. Figure 5.4 shows the shadow price for constraint 5.7. The objective function changes from $1303 to $1315 with a unit change in right hand side of the constraint. This shows that if the availability of D is increased, this will have a positive impact on profit. The shadow prices provide a way of improving the solution or to react appropriately when external circumstances create opportunities. Opportunity cost on the other hand is the rate of degradation of the optimum per unit use of non-basic(zero) variable in solution.

Solution Methods:

Simplex methods ([72, 71, 73]) move from boundary to boundary within the feasible region. The simplex methods requires initial basic solution to be feasible. There are various variants of simplex methods like dual simplex method, the Big M method, and the two-phase simplex method. Interior point methods on the other hand visit points within the interior of the feasible region more inline with the nonlinear programming methods. In general, good interior point methods perform as well or better than simplex codes on larger problems when no prior information about the solution is available. When such "warm start" information is available, simplex methods are able to make much better use of it than the interior point methods.

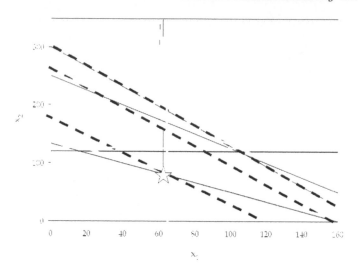

FIGURE 5.4
Shadow price of the constraint 5.7

5.4   Nonlinear Programming

In nonlinear programming (NLP) problems, either the objective function, the constraints, or both the objective and the constraints are nonlinear. Unlike LP, NLP solution does not always lie at the vertex of the feasible region. NLP optimum lies where the Jacobean of the function obtained by combining constraints with the objective function (using Lagrange multipliers as follows) is zero. The solution is local minimum if the Jacobian J is zero and the Hessian H is positive definite, and it is a local maximum if J is zero and H is negative definite.

$$L(x, \mu, \lambda) \;=\; z(x) \;+\; g(x)^T \mu \;+\; h(x)^T \lambda \qquad (5.10)$$

where λ and μ are *Lagrange multipliers*, also known as *dual variables* or *Kuhn-Tucker multipliers*. The first order optimality conditions are referred to as the Kuhn-Tucker conditions or Karush-Kuhn-Tucker (KKT) conditions and were developed independently by [74], and [75].

Kuhn-Tucker Conditions:

The first order Kuhn-Tucker conditions necessary for optimality can be written as follows:

   1.   Linear dependence of gradients:

$$\nabla L \;(x^*, \mu^*, \lambda^*) \;=\; \nabla Z(x^*) \;+\; \nabla g(x^*)^T \mu^* \;+\; \nabla h(x^*)^T \lambda^* \;=\; 0 \;\;(5.11)$$

where * refers to the optimum solution.

2. Feasibility of the NLP solution:

$$g(x^*) \leq 0 \tag{5.12}$$

$$h(x^*) = 0 \tag{5.13}$$

3. Complementarity condition; either $\mu^* = 0$ or $g(x^*) = 0$:

$$\mu^{*T} g(x^*) = 0 \tag{5.14}$$

4. Non-negativity of inequality constraint multipliers:

$$\mu^* \geq 0 \tag{5.15}$$

It should be remembered that the direction of inequality is very important here. The non-negativity requirement (for a minimization problem) above ensures that the constraint direction is not violated and the solution is in the feasible region. The sufficiency condition depends on Hessian as stated earlier.

Example 5.2: The following reaction is taking place in a batch reactor. $A \rightarrow R \rightarrow S$, where R is the desired product and A is the reactant, with an initial concentration, $CA_0 = 10$ moles/volume. The rate equations for this reaction are provided below. Solve the problem to obtain maximum concentration of R.

$$CA_t = CA_0 \exp{(-k_1 t)}$$

$$CR_t = CA_0 k_1 \left[\frac{\exp{(-k_1 t)}}{k_2 - k_1} - \frac{\exp{(-k_2 t)}}{k_2 - k_1} \right]$$

$$CS_t = CA_0 \left[1 - \frac{k_2 \exp{(-k_1 t)}}{k_2 - k_1} + \frac{k_1 \exp{(-k_2 t)}}{k_2 - k_1} \right]$$

where $k_1 = 10$ per hour, $k_2 = 1.0$ per hour

Solution: The problem is to maximize the concentration CR_t.

$$\text{Maximize } CR_t = CA_0 k_1 \left[\frac{\exp{(-k_1 t)}}{k_2 - k_1} - \frac{\exp{(-k_2 t)}}{k_2 - k_1} \right] \tag{5.16}$$
$$CA_t, CR_t, CS_t, t, x_2$$

subject to

$$CA_t = CA_0 \exp{(-k_1 t)} \tag{5.17}$$

$$CS_t = CA_0 \left[1 - \frac{k_2 \exp{(-k_1 t)}}{k_2 - k_1} + \frac{k_1 \exp{(-k_2 t)}}{k_2 - k_1} \right] \tag{5.18}$$

Converting the NLP into a minimization problem and formulating the augmented Lagrangian function results in the following unconstrained NLP.

$$Min \quad L = -CA_0 k_1 \left[\frac{\exp{(-k_1 t)}}{k_2 - k_1} - \frac{\exp{(-k_2 t)}}{k_2 - k_1} \right] + \lambda_1$$
$$CA_t, CS_t, t, \lambda_1, \lambda_2 \quad (CA_t - CA_0 \exp{(-k_1 t)}) + \lambda_2 (CS_t -$$
$$CA_0 \left[1 - \frac{k_2 \exp{(-k_1 t)}}{k_2 - k_1} + \frac{k_1 \exp{(-k_2 t)}}{k_2 - k_1} \right]) \tag{5.19}$$

$$\frac{\partial L}{\partial CA_t} = \lambda_1 = 0 \tag{5.20}$$

$$\frac{\partial L}{\partial CS_t} = \lambda_2 = 0 \tag{5.21}$$

$$\frac{\partial L}{\partial t} = -CA0k_1[\frac{-k_1\exp\left(-k_1t\right)}{k_2-k_1} - \frac{-k_2\exp\left(-k_2t\right)}{k_2-k_1}] = 0 \tag{5.22}$$

$$\frac{\partial L}{\partial \lambda_1} = CA_t - CA_0\exp\left(-k_1t\right) = 0 \tag{5.23}$$

$$\frac{\partial L}{\partial \lambda_2} = CS_t = CA_0[1 - \frac{k_2\exp\left(-k_1t\right)}{k_2-k_1} - \frac{k_1\exp\left(-k_2t\right)}{k_2-k_1}] = 0 \tag{5.24}$$

Solving the Equation 5.24 for t results in:

$$t_{opt} = \frac{\ln k_2/k_1}{k_2-k_1} = 0.2558 \ hrs \tag{5.25}$$

For the sufficiency condition.

$$\frac{\partial^2 L}{\partial t^2} = -CA0k_1[\frac{k_1{}^2\exp\left(-k_1t\right)}{k_2-k_1} - \frac{k_2{}^2\exp\left(-k_2t\right)}{k_2-k_1}] = 77.46 \tag{5.26}$$

The Hessian given by Equation 5.26 is positive, hence, it is a minimum. This is also clear from Figure 5.5.

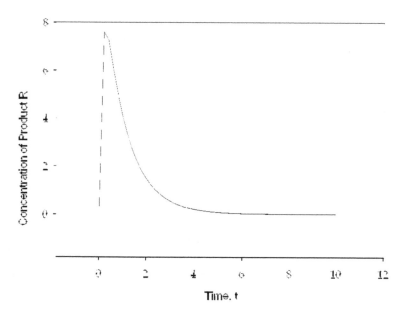

FIGURE 5.5
Concentration profile for product R

Solution Methods:

NLP solutions involve solving the set of nonlinear equations resulting from the KKT conditions. In nonlinear equation-solving procedures, the Newton-Raphson method shows the fastest convergence if one is away from the solution. Therefore, quasi-Newton methods are commonly used in solution methods for NLP. Currently, the two major methods for NLP commonly used in various commercial packages are: (1)the generalized reduced gradient (GRG) method and (2) the sequential quadratic programming (SQP). These two are quasi-Newton methods.

MINOS uses (also available in GAMS), a particular implementation of the GRG method. The basic idea behind the reduced gradient methods is to solve a sequence of subproblems with linearized constraints, where the subproblems are solved by variable elimination. SQP, on the other hand, takes the Newton step using the KKT conditions. It can be easily shown that the Newton step (if the exact Hessian is known) results in a quadratic programming problem (objective function quadratic, constraints linear) containing the Newton direction vector as decision variables. GRG methods are best suited for problems involving a significant number of linear constraints. On the other hand, SQP methods are useful for highly nonlinear problems. Extension of the successful interior-point methods for LPs to NLP is the subject of ongoing research.

5.5   Discrete Optimization

Discrete optimization problems involve discrete decision variables. Discrete optimization problems can be classified as integer programming (IP) problems, mixed integer linear programming (MILP), and mixed integer nonlinear programming (MINLP) problems.

Representation of the discrete decision space plays an important role in selecting a particular algorithm to solve the discrete optimization problem. For representing discrete decisions one can assign an integer for each shape or a binary variable having values of 0 and 1 (1 corresponding to yes and 0 to no). The binary variable representation is used in traditional mathematical programming algorithms for solving problems involving discrete decision variables. However, probabilistic methods such as simulated annealing and genetic algorithms which are based on analogies to a physical process such as the annealing of metals or to a natural process such as genetic evolution, may prefer to use different integers assigned to different decisions.

The commonly used mathematical programming method for solving IP, MILP, and MINLP is the branch and bound method which uses a tree structure to define the problem and the associated binary variables. The following example shows the tree structure for the problem.

Example 5.3: Given a mixture of four components A, B, C, D for which separation technologies given in Table 5.1 are to be considered.
Determine the tree for all the alternative sequences.

Solution: Figure 5.6 shows the tree structure for problem formulation in thousands of dollars.

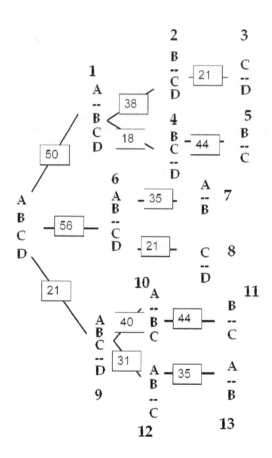

FIGURE 5.6
Tree representation for branch and bound

TABLE 5.1

Cost of separators in $/year

Separator	Cost
A/BCD	50,000
AB/CD	56,000
ABC/D	21,000
A/BC	40,000
AB/C	31,000
B/CD	38,000
BC/D	18,000
A/B	35,000
B/C	44,000
C/D	21,000

Having developed the representation, the question is how to search for the optimum. One can go through the complete enumeration, but that would involve evaluating each node of the tree. The intelligent way is to reduce the search space by implicit enumeration and evaluate as few nodes as possible. Consider the above example of separation sequencing. The objective is to minimize the cost of separation. If one looks at the nodes for each branch, there are an initial node, intermediate nodes, and a terminal node. Each node is the sum of the costs of all earlier nodes in that branch. Because this cost increases monotonically as we progress through the initial, intermediate, and final nodes, we can define the upper bound and lower bounds for each branch.

- The cost accumulated at any intermediate node is a lower bound to the cost of any successor nodes, as the successor node is bound to incur additional cost.

- For a terminal node, the total cost provides an upper bound to the original problem because a terminal node represents a solution that may or may not be optimal.

These heuristics allow us to prune the tree. If the cost at the current node is greater than or equal to the upper bound defined earlier either from one of the prior branches or known to us from experience, then we don't need to go further in that branch. These are the two common ways to prune the tree based on the order in which the nodes are enumerated:

- Depth-first: Here, we successively perform one branching on the most recently created node. When no nodes can be expanded, we backtrack to a node whose successor nodes have not been examined.

- Breadth-first: Here, we select the node with the lowest cost and expand all its successor nodes.

Example 5.4: Consider the tree structure in Figure 5.6. Use breadth-first strategy to prune the tree.

Solution: Figure 5.6 shows the tree structure and cost of each node. The nodes enumerated are shown by dark lines. It can be seen that out of 13 only 12 nodes (node 3 is not evaluated) need to be evaluated by this strategy. In general breadth-first strategy requires fewer nodes to be examined but requires higher memory storage than the depth-first strategy. Since storage was a problem for breadth-first generation computers depth-first strategy is commonly used.

Numerical Methods for IP, MILP, MINLP:

In Example 5.4, we could carry out the branch and bound method using graphical representation. In real practice, for numerical computation we need algebraic representation of the tree. The following example shows the algebraic representation in terms of binary variables.

Example 5.5: Provide the algebraic representation of the problem specified in Example 5.3 for the numerical branch-and-bound procedure.

Solution: Consider the tree representation for this problem as shown in Figure 5.6. As stated earlier, the decision variables associated with each node can be represented by binary variables y_i. The logic is that if the node were present in the sequence, the binary variable associated with that node would be equal to one, else it would be zero. C_i denotes the cost of each node, and y_i represents the binary variable associated with each node. They are numbered (as subscripts) according to the nodes shown in the figure (e.g., y_9 corresponds to Node 9). Let us translate the tree structure into logical constraints. The objective function is the minimization of total costs, the cost of each node present in the final sequence. Because we do not know which node will be selected, we can write the objective function in terms of the cost of each node multiplied by the binary variable. Given that node not appearing in the sequence, the corresponding binary variable y will go to zero and will not contribute to the objective function.

$$\text{Min} \quad z = \sum_{i=1}^{13} C_i y_i$$
$$y_i$$

subject to:
At the Root Node we can only select one of the three nodes.

$$y_1 + y_6 + y_9 = 1$$

Node 2 or Node 4 will exist if Node 1 is considered.

$$y_2 + y_4 = y_1$$

Node 3 will exist if Node 2 is considered.

$$y_3 = y_2$$

Node 5 will exist if Node 4 is considered.

$$y_5 = y_4$$

Node 7 will exist if Node 6 is present and Node 8 will exist only if Node 7 is considered.

$$y_7 = y_6$$

$$y_8 = y_7$$

Node 10 or Node 12 will exist if Node 9 is considered.

$$y_{10} + y_{12} = y_9$$

Node 11 will exist if Node 10 is present and Node 13 will exist if Node 12 is considered.

$$y_{11} = y_{10}$$

$$y_{13} = y_{12}$$

It is obvious from the above example that once the discrete variables are assigned, it is possible to write logical constraints. Typical examples are:

1. Multiple Choice Constraints:

    •Select only one item:
    $$\sum_i y_i = 1$$

    •Select at most one item:
    $$\sum_i y_i \leq 1$$

    •Select at least one item:
    $$\sum_i y_i \geq 1$$

2. Implication Constraints:

    •If item k is selected, item j must be selected, but not necessarily vice versa:
    $$y_k - y_j \leq 0$$

    •If binary variable y is zero, an associated continuous variable x must be zero:

    $$x - Uy \leq 0$$
    $$x \geq 0$$

    where U is an upper limit to x.

3. Either-or constraints:

•Either constraint $g_1(x) \geq 0$ or constraint $g_2(x) \geq 0$ must hold:

$$g_1(x) \; - \; Uy \; \leq \; 0$$

$$g_2(x) \; - \; U(1-y) \; \leq \; 0$$

where U is a large value.

As can be seen above, the IP problem can be represented by the following generalized form.

$$\text{Optimize} \quad Z \; = \; z(y) \; = \; \sum_i c_i y_i \; = \; C^T y \qquad (5.27)$$

$$y_i$$

subject to

$$h(y) \; = \; A^T y \; + B \; = \; 0 \qquad (5.28)$$

$$g(y) \; = \; D^T y \; + \; E \; \leq \; 0 \qquad (5.29)$$

where $y_i \; \epsilon \; 0,1$.

As can be seen above IPs tend to be linear. The easiest way to solve this is to relax the IP into LP and use that as a starting point for the branch and bound method.

The mixed integer linear programming problems are of the form given below:

$$\text{Optimize} \quad Z \; = \; z(x,y) \; = \; a^T y \; + \; C^T x \qquad (5.30)$$

$$x, y_i$$

where $y_i \; \epsilon \; 0,1$ and x is a set of continuous variables. Note that the IP part in the objective function is again linear.

Subject to

$$g(x,y) \; = \; - By \; + \; A^T x \; \leq \; 0 \qquad (5.31)$$

Branch-and-bound is a commonly used technique for solving MILP problems, where at each node, instead of looking at the fixed costs as we have seen in Figure 5.6, an LP is solved.

The following is the generalized representation of an MINLP problem.

$$\text{Optimize} \quad Z \; = \; z(x,y) \; = \; a^T y \; + \; f(x) \qquad (5.32)$$

$$x, y_i$$

where $y_i \; \epsilon \; 0,1$ and x is a set of continuous variables. The first term represents a linear function involving the binary variables y and the second term is a nonlinear function in x. This formulation avoids nonconvexities and bilinear terms in the objective function.

Similarly, for the constraints the following formulation is used.

subject to

$$h(x) \; = \; 0 \qquad (5.33)$$

$$g(x,y) \; = \; - B^T y \; + \; g(x) \; \leq \; 0 \qquad (5.34)$$

The branch-and-bound approach to MINLP can encounter problems of singularities and infeasibilities, and can be computationally expensive. Generalized Bender's Decomposition (GBD)[76] and Outer Approximation (OA)[77] algorithms tend to be more efficient than branch-and-bound, and are commonly employed to solve MINLPs. GAMS uses OA for MINLP solution. However, these algorithms are designed for open equation systems and encounter difficulties when functions do not satisfy convexity conditions, for systems having a large combinatorial explosion, or when the solution space is discontinuous. Probabilistic methods such as simulated annealing and genetic algorithms provide an alternative to these algorithms.

Simulated Annealing and Genetic Algorithms:

Simulated annealing (SA) and genetic algorithms (GA) are combinatorial methods based on ideas from the physical world.

Simulated annealing is a heuristic based combinatorial optimization method based on ideas from statistical mechanics (Kirkpatrick et al., 1983). The analogy is to the behavior of physical systems in the presence of a heat bath: in physical annealing, all atomic particles arrange themselves in a lattice formation that minimizes the amount of energy in the substance, provided the initial temperature is sufficiently high and the cooling is carried out slowly. At each temperature T, the system is allowed to reach thermal equilibrium, which is characterized by the probability (P_r) of being in a state with energy E given by the Boltzmann distribution:

$$P_r(Energy\ state\ =\ E)\ =\ \frac{1}{\phi(t)}\exp\left(-\frac{E}{K_bT}\right) \tag{5.35}$$

where K_b is Boltzmann's constant $(1.3806 \times 10^{23}\ J/K)$ and $1/\phi(t)$ is a normalization factor [78].

In SA, the objective function (usually cost) becomes the energy of the system. The goal is to minimize the cost (energy). Simulating the behavior of the system then becomes a question of generating a random perturbation that displaces a "particle" (moving the system to another configuration). If the configuration that results from the move has a lower energy state, the move is accepted. However, if the move is to a higher energy state, the move is accepted according to the Metropolis criteria (accepted with probability $= \exp\left(-\Delta E/K_bT\right)$;[79]).

This implies that at high temperatures, a large percentage of uphill moves is accepted. However, as the temperature gets colder, a small percentage of uphill moves is accepted. After the system has evolved to thermal equilibrium at a given temperature, the temperature is lowered and the annealing process continues until the system reaches a temperature that represents "freezing." Thus, SA combines both iterative improvements in local areas and random jumping to help ensure that the system does not get stuck in a local optimum. The general procedure for SA is as follows (VanLaarhoven and Aarts, 1987).

1. Get an initial solution configuration S.
2. Get an initial temperature, $T = T_{initial}$.
3. While not yet frozen $(T > T_{froze})$ perform the following.
   (a) Perform the following loop K times until equilibrium is reached (K is the number of the moves per temperature level and is a function of moves accepted at that temperature level).

- Generate a move S' by perturbing S.
- Let $\Delta = Cost(S') - Cost(S)$.
- If $\Delta \leq 0$ (accept downhill move for minimization), then set $S = S'$ (accept the move), else, if $\Delta > 0$, it is an uphill move, accept the move with probability $\exp(-\Delta/T)$.
- Update number of accepts and rejects.
- Determine K and return to Step (a).

   (b) No significant change in last C steps. Go to Step (4).

   (c) Decrease T and go to Step (3).

  4. Optimum solution is reached.

Genetic algorithms are search algorithms based on the mechanics of natural selection and natural genetics. Based on the idea of survival of the fittest, they combine the fittest string structures with a structured yet randomized information exchange to form a search algorithm with some of the innovative flair of human search [80].

A GA is a search procedure modeled on the mechanics of natural selection rather than a simulated reasoning process. Domain knowledge is embedded in the abstract representation of a candidate solution, termed an organism, and organisms are grouped into sets called populations. Successive populations are called generations. A general GA creates an initial generation (a population or a discrete set of decision variables) $G(0)$, and for each generation $G(t)$, generates a new one $G(t+1)$. The general procedure for GA is described below.

At $t = 0$,

- Generate initial population randomly, $G(t)$.
- Evaluate $G(t)$. Find fitter solutions.
- While termination criteria are not satisfied, do

   $t = t + 1$.

- Select $G(t)$.
- Recombine $G(t)$ using reproduction, crossover and mutation, and immigration strategies.
- Evaluate $G(t)$. Find the fitter solutions.
- Repeat the loop until solution is found.

5.6   Optimal Control

The unsteady state nature of batch processes results in differential algebraic optimization (DAOP) problems. Underlying a DAOP is the problem of optimal control where time dependent decisions are decided. A differential algebraic optimization problem in general can be stated as follows.

$$\text{Optimize} \quad J = j(\overline{x}_T) + \int_0^T k(\overline{x}_t, \theta_t, x_s) \, dt \tag{5.36}$$

$$\theta_t, x_s$$

subject to

$$\frac{d\overline{x}_t}{dt} = f(\overline{x}_t, \theta_t, x_s) \tag{5.37}$$

$$h(\overline{x}_t, \theta_t, x_s) = 0 \tag{5.38}$$

$$g(\overline{x}_t, \theta_t, x_s) \leq 0 \tag{5.39}$$

$$\overline{x}_0 = \overline{x}_{initial}$$

$$\theta(L) \leq \theta_t \leq \theta(U)$$

$$x_s(L) \leq x_s \leq x_s(U)$$

where J is the objective function given by Equation 5.36, $\overline{x}_t$ is the state variable vector ($nx \times 1$ dimensional) at any time t, θ_t is the control vector, and x_s represents the scalar variables. It is obvious that the objective function can only be calculated at the end of operation T. Equations 5.38 and 5.39 represent the equality (m_1 constraints) and inequality constraints (m_2 constraints, including bounds on the state variables), respectively (constituting a total of m constraints). $\theta(L)$ and $x_s(L)$ represent the lower bounds on the set of control variables θ_t and the scalar variable x_s, respectively, and $\theta(U)$, $x_s(U)$ are the corresponding upper bounds. In the absence of the scalar decision variables x_s, a DAOP is equivalent to an optimal control problem. In general mathematical methods to solve optimal control problems involve calculus of variations, the maximum principle and the dynamic programming technique. NLP techniques can also be used to solve this problem provided all the system of differential equations are converted to nonlinear algebraic equations.

Calculus of Variations:

Calculus of variations considers the entire path of the function and optimizes the integral by minimizing the functional by making the first derivative vanish (first-order condition for nonlinear systems), resulting in second-order differential equations given by Euler-Lagrangian equations. In this formulation, the objective function is augmented to include constraints through the use of Lagrangian multipliers μ_i and λ_j, as given below.

$$\text{Optimize} \quad L - j(\overline{x}_T) \int_0^T k(\overline{x}_t, \theta_t, x_s) \, dt$$

$$\theta_t, \mu_i, \lambda_{jj}, \lambda_{kk} \quad + \sum_i \mu_i^T \left(\frac{dx_i}{dt} - f(\overline{x}_t, \theta_t, x_s) \right) +$$

$$+ \sum_{jj=1}^{JJ} \lambda_{jj} h(\overline{x}_t, \theta_t, x_s)$$

$$+ \sum_{kk=JJ+1}^{KK} \lambda_{kk} (g(\overline{x}_t, \theta_t, x_s) \tag{5.40}$$

By applying the first-order condition for optimization, that is, the first derivative with respect to the control variable, and Lagrange multipliers should disappear resulting in Euler-Lagrangian differential equations given below

$$\frac{\partial L}{\partial \theta} - \frac{d(\frac{\partial L}{\partial \theta'})}{dt} = 0 \tag{5.41}$$

$$\frac{\partial L}{\partial x_i} - \frac{d(\frac{\partial L}{\partial x_i'})}{dt} = 0 \tag{5.42}$$

$$h(\overline{x}_t, \theta_t, x_s) = 0 \tag{5.43}$$

$$g_l(\overline{x}_t, \theta_t, x_s) = 0 \; \lambda_l \geq 0 \tag{5.44}$$

$$g_m(\overline{x}_t, \theta_t, x_s) \leq 0 \; \lambda_m = 0 \tag{5.45}$$

where $\theta' = d\theta/dt$ and $x_i' = dx_i/dt$.

The following example of batch distillation optimal control (optimal reflux policy problem illustrates this.

The maximum distillate problem in batch distillation is described as- where the amount of distillate of a specified concentration for a specified time is maximized. [47, 48, 49, 50, 51])

Example 5.6: The maximum distillate problem for separation of a binary mixture involves solution of the following problem:

$$\text{Maximize} \quad J = \int_0^T \frac{dD}{dt} \, dt = \int_0^T \frac{V}{R_t + 1} \, dt, \tag{5.46}$$
$$R_t$$

subject to the following purity constraint on the distillate

$$x_{Dav} = \frac{\int_0^T x_D^{(1)} \frac{V}{R_t + 1} \, dt}{\int_0^T \frac{V}{R_t + 1} \, dt} = x_D^* \tag{5.47}$$

and the following differential equations governing the dynamics of the column. The rest of the column is assumed to be at quasi-steady state and to obey the plate-to-plate or the shortcut method calculation equations.

$$\frac{dx_t^1}{dt} = \frac{dB_t}{dt} = \frac{-V}{R_t + 1}, \quad x_0^1 = B_0 = F, \tag{5.48}$$

$$\frac{dx_t^2}{dt} = \frac{dx_B^{(1)}}{dt} = \frac{V}{R_t + 1} \frac{(x_t^2 - x_D^{(1)})}{x_t^1}, \quad x_0^2 = x_F^{(1)} \tag{5.49}$$

Formulate the maximum distillate problem using the calculus of variations.

Solution: Since this problem contains equality constraints, we need to use the Euler-Lagrangian formulation. First, all three equality constraints (Equations 5.47 to 5.49) are augmented to the objective function to form a new objective function

L given by:

Maximize $\quad L = \displaystyle\int_0^T \frac{V}{R_t + 1}\, dt\, \left[1 - \lambda(x_D^* - x_D^{(1)})\right] -$

$x_1,\ x_2,\ R_t$

$$\mu_1\left[\frac{dB_t}{dt} - \frac{-V}{R_t + 1}\right] - \mu_2\left[\frac{dx_t^2}{dt} - \frac{V}{R_t + 1}\frac{(x_B^{(1)} - x_D^{(1)})}{x_t^1}\right]$$

$$(5.50)$$

where λ is a scalar Lagrange multiplier and μ_i, $i = 1, 2$ are the Lagrangian multipliers as a function of time. Application of Euler-Lagrangian differential equations leads to the following three Euler-Lagrange equations.

$$\frac{\partial L}{\partial x_t^1} \quad - \quad \frac{d(\frac{\partial L}{\partial x_t^{1'}})}{dt} \;=\; 0$$

$$\frac{d\mu_1}{dt} \;=\; \mu_2\left[\frac{V}{R_t + 1}\frac{(x_t^2 - x_D^{(1)})}{x_t^{1\,2}}\right] \tag{5.51}$$

$$\frac{\partial L}{\partial x_t^2} \quad - \quad \frac{d(\frac{\partial L}{\partial x_t^{2'}})}{dt} \;=\; 0$$

$$\frac{d\mu_2}{dt} \;=\; -\frac{V}{R_t + 1}\lambda\left(\frac{\partial x_D^{(1)}}{\partial x_t^2}\right)_{R_t} - \mu_2\frac{V}{x_t^1(R_t + 1)}\left[1 - \left(\frac{\partial x_D^{(1)}}{\partial x_t^2}\right)_{R_t}\right] \tag{5.52}$$

$$\frac{\partial L}{\partial R_t} \quad - \quad \frac{d(\frac{\partial L}{\partial R_t'})}{dt} \;=\; 0$$

$$R_t \;=\; \frac{\left[\frac{\mu_2}{x_t^1}(x_t^2 - x_D^{(1)}) - \mu_1 - \lambda(x_D^* - x_D^{(1)}) + 1\right]}{\frac{\partial x_D^{(1)}}{\partial R_t}\left(\lambda - \frac{\mu_2}{x_t^1}\right)} - 1 \tag{5.53}$$

Maximum Principle

In the maximum principle [81, 82] formulation (the right-hand side of Equation 5.54), the objective function is represented as a linear function in terms of the final values of $\overline{x}$ and the values of $\overline{c}$, where $\overline{c}$ represents the vector of constants. The maximum principle formulation for the above-mentioned DAOP is given below:

Maximize $\quad J = j(\overline{x}_T) + \displaystyle\int_0^T k(\overline{x}_t, \theta_t, x_s)dt = \overline{c}^T\overline{x}_T = \sum_{i=1}^{nx} c_i x_i(T)$

θ_t

$$(5.54)$$

subject to

$$\frac{d\overline{x}_t}{dt} \;=\; f(\overline{x}_t, \theta_t, x_s) \tag{5.55}$$

$$h(\bar{x}_t, \theta_t, x_s) = 0 \tag{5.56}$$

$$g(\bar{x}_t, \theta_t, x_s) \leq 0 \tag{5.57}$$

$$\bar{x}_0 = \bar{x}_{initial}$$

By using the Lagrangian formulation for the above problem, fixing scalar variables x_s, and removing the bounds $\theta(L)$ and $\theta(U)$ on the control variable vector θ_t, one obtains:

$$\text{Maximize} \quad J^* = \bar{c}^T \bar{x}_T + \overline{\lambda_1}(h(\bar{x}_t, \theta_t, x_s)) + \overline{\lambda_2}(g(\bar{x}_t, \theta_t, x_s))$$
$$\theta_t \tag{5.58}$$

subject to

$$\frac{d\bar{x}_t}{dt} = f(\bar{x}_t, \theta_t, x_s) \tag{5.59}$$

$$\bar{x}_0 = \bar{x}_{initial}$$

where

$$\bar{\lambda} = [\overline{\lambda_1}, \overline{\lambda_2}]$$

Application of the maximum principle to the above problem involves the addition of nx adjoint variables z_t (one adjoint variable per state variable), nx adjoint equations, and a Hamiltonian, which satisfies the following relations:

$$H(\bar{z}_t, \bar{x}_t, \theta_t) = \bar{z}_t^T f(\bar{x}_t, \theta_t, x_s) = \sum_{i=1}^{nx} z_i f_i(\bar{x}_t, \theta_t) \tag{5.60}$$

$$\frac{dz_i}{dt} = -\sum_{j=1}^{nx} z_j \frac{\partial f_j}{\partial x_i} \tag{5.61}$$

$$\bar{z}_T = \bar{c} \tag{5.62}$$

The boundary conditions given above (Equation 5.62) are often true, but not always. When present, they play an important role in the final stages of the solution. Therefore, it is important to keep track of the boundary conditions. As stated earlier, we have one objective H for each time step. The optimal decision vector θ_t can be obtained by extremizing the Hamiltonian given by Equation 5.60 for each time step. θ_t can then be expressed as:

$$\theta_t = H^*(\bar{x}_t, \bar{z}_t, \bar{\lambda}) \tag{5.63}$$

where H* denotes the function obtained by using the stationary condition $(dH_t/d\theta_t)$ for the Hamiltonian.

The following maximum distillate problem illustrates the principle.

Example 5.7: Formulate the maximum distillate problem presented in Example 5.6 using the maximum principle and compare the formulation with that obtained using variational calculus (solution for Example 5.6).

Solution: The maximum distillate problem can be written as

$$\text{Maximize} \quad J = \int_0^T \frac{dD}{dt} \, dt = \int_0^T \frac{V}{R_t + 1} \, dt, \tag{5.64}$$
$$R_t$$

subject to the following purity constraint on the distillate

$$x_{Dav} = \frac{\int_0^T x_D^{(1)} \frac{V}{R_t + 1} \, dt}{\int_0^T \frac{V}{R_t + 1} \, dt} = x_D^* \tag{5.65}$$

The constraint on the purity is removed by employing the method of Lagrange multipliers. By combining Equations 5.64 and 5.65:

$$\underset{R_t}{\text{Maximize}} \quad L = \int_0^T \frac{V}{R_t + 1} \left[1 - \lambda(x_D^* - x_D^{(1)}) \right] \, dt, \tag{5.66}$$

where λ is a Lagrange multiplier. Now the objective function is to maximize L, instead of J. To solve this problem, an additional state variable $x_{3,t}$ is introduced, which is given by

$$x_{3,t} = \int_0^t \frac{V}{R_t + 1} \left[1 - \lambda(x_D^* - x_D^{(1)}) \right] \, dt, \tag{5.67}$$

The problem can then be rewritten as

$$\underset{R_t}{\text{Maximize}} \quad x_{3,T}, \tag{5.68}$$

subject to the following differential equations for the three state variables and the time-implicit model for the rest of the column,

$$\frac{dx_{1,t}}{dt} = \frac{-V}{R_t + 1}, \quad x_{1,0} = B_0 = F, \tag{5.69}$$

$$\frac{dx_{2,t}}{dt} = \frac{V}{R_t + 1} \frac{(x_{2,t} - x_D^{(1)})}{x_{1,t}}, \quad x_{2,t} = x_F^{(1)} \tag{5.70}$$

$$\frac{dx_{3,t}}{dt} = \frac{V}{R_t + 1} \left[1 - \lambda(x_D^* - x_D^{(1)}) \right] \, dt, \tag{5.71}$$

The Hamiltonian function, which should be maximized, is:

$$H_t = -z_{1,t} \frac{V}{R_t + 1} + z_{2,t} \frac{V(x_{2,t} - x_D^{(1)})}{(R_t + 1)x_{1,t}} + z_{3,t} \frac{V}{R_t + 1} \left[1 - \lambda(x_D^* - x_D^{(1)}) \right] \tag{5.72}$$

and the adjoint equations are:

$$\frac{dz_{1,t}}{dt} = z_{2,t} \frac{V(x_{2,t} - x_D^{(1)})}{(R_t + 1)(x_{1,t})^2}, \quad z_{1,T} = 0, \tag{5.73}$$

$$\frac{dz_{2,t}}{dt} = -z_{2,t} \frac{V(1 - \frac{\partial x_D^{(1)}}{\partial x_{2,1}})}{(R_t + 1)x_{1,t}} - z_{3,t} \lambda \frac{V}{(R_t + 1)} (\frac{\partial x_D^{(1)}}{\partial x_{2,t}}), \quad z_{2,T} = 0 \tag{5.74}$$

and

$$\frac{dz_{3,t}}{dt} = 0, \quad z_{3,T} = 1 \tag{5.75}$$

Since the above equation for z_t^3 gives

$$z_{3,t} = 1, \qquad (5.76)$$

the Hamiltonian function in Equation 5.72 can be written as

$$H_t = -z_{1,t} \frac{V}{R_t + 1} + z_{2,t} \frac{V(x_{2,t} - x_D^{(1)})}{(R_t + 1)x_{1,t}} + \frac{V}{R_t + 1}\left[1 - \lambda(x_D^* - x_D^{(1)})\right] \qquad (5.77)$$

and

$$\frac{dz_{2,t}}{dt} = -z_{2,t} \frac{V\left(1 - \frac{\partial x_D^{(1)}}{\partial x_{t,2}}\right)}{(R_t + 1)x_{1,t}} - \lambda \frac{V}{(R_t + 1)}\left(\frac{\partial x_D^{(1)}}{\partial x_{2,t}}\right), \quad z_{2,T} = 0 \qquad (5.78)$$

From the optimality condition $\partial H/\partial R_t = 0$, it follows that

$$R_t = \frac{\left[\frac{z_{2,t}}{x_{1,t}}(x_{2,t} - x_D^{(1)}) - z_{1,t} - \lambda(x_D^* - x_D^{(1)}) + 1\right]}{\frac{\partial x_D^{(1)}}{\partial R_t}\left(\lambda - \frac{z_{2,t}}{x_{1,t}}\right)} - 1 \qquad (5.79)$$

It can be easily seen from Equation 5.53 in Example 5.6 and from Equation 5.79 that the two formulations lead to the same results where in the case of variational calculus the time-dependent Lagrange multipliers μ_i are equivalent to the adjoint variables z_i in the maximum principle formulation.

Dynamic Programming

The method of dynamic programming is based on the principle of optimality, as stated by [83].

An optimal policy has the property that whatever the initial state and initial decision are the remaining decisions must constitute an optimal policy with regard to the state resulting from the first decision.

In short, the principle of optimality states that the minimum value of a function is a function of the initial state and the initial time and results in Hamilton-Jacobi-Bellman equations (H-J-B) given below.

$$0 = \left[\frac{\partial J}{\partial t}\right] + H\left(\overline{x}(t), \frac{\partial J}{\partial \overline{x}}, t\right) \qquad (5.80)$$

where

$$H = \underset{\theta_T}{\text{Optimize}} \quad k(\overline{x}_T, \theta_T) + \left[\frac{\partial J}{\partial \overline{x}}\right]^\tau [f(\overline{x}_T, \theta_T, \overline{\mu}] \qquad (5.81)$$

With the boundary conditions:

$$\frac{\partial J}{\partial x(0)} = 0 \qquad (5.82)$$

$$\left[\frac{\partial J}{\partial t}\right]^T = H\left(\overline{x}(T), \frac{\partial J}{\partial \overline{x}}, T\right) \qquad (5.83)$$

As can be seen, the H-J-B equation are partial differential equations. The following example illustrates the use of H-J-B equations. [84] used the method of characteristics, described below for the above equations. In the method of characteristics a dummy variable s is introduced which converts the above equations into the series of equations given below.

$$\frac{dT}{ds} = 1, \quad T_{initial} = 0 \tag{5.84}$$

$$\frac{d\bar{x}(T)}{ds} = f(\bar{x}_T, \theta_T, \bar{\mu}) \tag{5.85}$$

$$\frac{dJ}{ds} = k(\bar{x}_T, \theta_T) \tag{5.86}$$

$$\frac{d\left(\frac{\partial J}{dx_{i,T}}\right)}{ds} = J(x_{i(T)}, T) - \left(\frac{\partial J}{\partial x_{i,T}}\right) f(x_{i,T}, \theta_T, \bar{\mu})$$

$$\left(\frac{\partial J}{dx_{i,T}}\right)_0 = 0 \tag{5.87}$$

$$\frac{d\left(\frac{\partial J}{\partial T}\right)}{ds} = 0$$

$$\left(\frac{\partial J}{\partial T}\right)_0 = 0 \tag{5.88}$$

These equations along with the boundary conditions need to be solved to get the optimal control policy.

Example 5.8: Solve the maximum distillate problem presented in Example 5.6 using the dynamic programming technique and compare the formulation with that obtained using variational calculus (solution for Example 5.6) and maximum principle (solution for Example 5.7).

Solution: Once again the Lagrange multiplier formulation of the maximum distillate problem can be written as

$$\underset{R_t}{\text{Maximize}} \quad L = \int_0^T \frac{V}{R_t + 1}\left[1 - \lambda(x_D^* - x_D^{(1)})\right] dt, \tag{5.89}$$

subject to:

$$\frac{dx_{1,t}}{dt} = \frac{-V}{R_t + 1}, \quad x_{1,0} - D_0 = F, \tag{5.90}$$

$$\frac{dx_{2,t}}{dt} = \frac{V}{R_t + 1}\frac{(x_{2,t} - x_D^{(1)})}{x_{1,t}}, \quad x_{2,0} = x_F^{(1)} \tag{5.91}$$

The Hamilton-Jacobi-Bellman equation for this problem can be written as:

$$\frac{\partial L}{\partial t} + \underset{R_T}{\text{Optimize}} \quad \frac{V}{R_T + 1}\frac{(x_{2,T} - x_D^{(1)})}{x_{1,T}}\frac{\partial L}{\partial x_{2,T}} + \frac{-V}{R_t + 1}\frac{\partial L}{\partial x_{1,T}} = 0 \tag{5.92}$$

The above quasi-linear partial differential equation can be solved by using equations based on the method of characteristics (Equations 5.84-5.88).

$$\frac{dT}{ds} = 1, T_{initial} = 0 \tag{5.93}$$

$$\frac{dx_{1,T}}{ds} = \frac{-V}{R_T + 1} \tag{5.94}$$

$$\frac{dx_{2,T}}{ds} = \frac{V}{R_T + 1} \frac{(x_{2,T} - x_D^{(1)})}{x_T^1} \tag{5.95}$$

$$\frac{dL}{ds} = \frac{V}{R_T + 1} \left[1 + \lambda(x_D^{(1)} - x_D^*) \right] \tag{5.96}$$

$$\frac{d\left(\frac{\partial L}{\partial x_{1,T}}\right)}{ds} = \frac{-V}{R_T + 1} \frac{(x_{2,T} - x_D^{(1)})}{(x_{1,T})^2} \quad (\frac{\partial L}{\partial x_{1,T}})_0 = 0 \tag{5.97}$$

$$\frac{d\left(\frac{\partial L}{\partial x_{2,T}}\right)}{ds} = \frac{V}{R_T + 1} \frac{\left(1 - \frac{\partial x_D^{(1)}}{\partial x_{2,T}}\right)}{(x_{1,T})^2} \left(\frac{\partial L}{\partial x_{2,T}}\right) + \frac{V}{R_T + 1} \left(\frac{\partial x_D^{(1)}}{\partial x_{2,T}}\right)_{R_T}$$

$$(\frac{\partial L}{\partial x - 2, T})_0 = 0 \tag{5.98}$$

$$\frac{d\left(\frac{\partial L}{\partial T}\right)}{ds} = 0$$

$$(\frac{\partial L}{\partial T})_0 = H \tag{5.99}$$

$$H = \text{Optimize} \quad \frac{V}{R_T + 1} \frac{(x_{2,T} - x_D^{(1)})}{x_{1,T}} \frac{\partial L}{\partial x_{2,T}} + \frac{-V}{R_T + 1} \frac{\partial L}{\partial x_{1,T}}$$

$$R_T$$

A differential equation for dR_T/ds could have been used along with differential Equations 5.93 to 5.98 to obtain the optimal policy. However, an explicit equation can be obtained by differentiating the right hand side of Equation 5.99 with respect to R_T and setting it to zero as given below.

$$R_T = \frac{\left[\frac{\partial L/\partial x_{2,T}}{x_{1,T}}(x_{2,T} - x_D^{(1)}) - \partial L/\partial x_{1,T} - \lambda(x_D^* - x_D^{(1)}) + 1 \right]}{\frac{\partial x_D^{(1)}}{\partial R_T}\left(\lambda - \frac{\partial L/\partial x_{2,T}}{x_{1,T}}\right)} - 1 \tag{5.100}$$

It can be easily seen from Equation 5.53 in Example 5.6 and Equation 5.79 in Example 5.7 as well as in the above equation that the formulations using dynamic programming, the maximum principle, and the calculus of variations lead to the same results, where in the case of variational calculus the time-dependent Lagrange multipliers μ_i are equivalent to the adjoint variables z_i in the maximum principle formulation, which are equivalent to the partial derivatives of the function L with respect to the state variables in the dynamic programming formulation.

5.7 Summary

Optimization problems can be divided into various categories such as LP, NLP, and MINLP depending on the type of objective function, constraints, and/or decision variables. Nonlinear programming problems involve either the objective function or the constraints, or both the objective function and the constraints are nonlinear. NLP can have multiple local optima. The NLP local optimum is a global minimum if the feasible region and the objective function are convex, and is a global maximum if the feasible region is convex and the objective function is concave. Karush-Kuhn-Tucker conditions provide necessary conditions for an NLP solution and are used in quasi-Newton methods to solve the problem iteratively. Discrete optimization involves integer programming (IP), mixed integer linear programming (MILP), and mixed integer nonlinear programming (MINLP) problems. The commonly used solution method for solving IP and MILP problems is the branch-and-bound method. Probabilistic methods such as simulated annealing and genetic algorithms provide an alternative to the branch-and-bound method. These methods are also considered global optimization methods. An optimal control problem involves vector decision variables. These problems are a subset of differential algebraic optimization problems. If the underlying differential equations can be discretized into a set of algebraic equations, then these problems can be solved using traditional NLP techniques. Otherwise, one has to resort to either the calculus of variations, the maximum principle, or the dynamic programming approach.

Notations

$\frac{dB}{dt}$	bottom rate [mol/time]
C_A	concentration of A [mol/vol]
C_{A0}	concentration of A at time=0 [mol/length^3]
C_R	concentration of R [mol/length^3]
C_S	concentration of S [mol/length^3]
$\frac{dD}{dt}$	distillate rate [mol/time]
E	energy term in simulated annealing
F	amount of feed [mol]
g	inequality constraint
h	equality constraint
H	Hamiltonian
k_1	reaction constant
k_2	reaction constant
K_b	Boltzmann's constant
L	Lagrangian function
n	number of components
N	number of plates
P_r	probability term in simulated annealing

R	reflux ratio
R_t	reflux ratio as a function of time
t	integration time [time]
T	batch time for batch distillation [time]
	temperature for simulated annealing
x	continuous decision variable
x_B	liquid-phase mole fraction of the reboiler
x_D	liquid-phase mole fraction of the distillate
$x_{D,avg}$	average distillate mole fraction
x_F	liquid-phase mole fraction of the feed
y	binary decision variable
$Z(or z)$	objective function
z_i	i-th adjoint variables in maximum principle formulation

Greek Letters:

α	relative volatility
θ	control variable
λ	Lagrangian multiplier for equality constraint
μ	Lagrangian multiplier for inequality constraint
ϕ	normalization factor in simulated annealing

6

Batch Absorption

CONTENTS

In batch gas absorption, a soluble vapor is absorbed from a mixture by means of a liquid in which the gas is more or less soluble. Unlike distillation, absorption is based on different solubility of the components in liquid and not based on boiling point difference. Absorption followed by reaction in the liquid phase is often used to get more complete removal of a solute from a gas mixture. Although, most of the absorption columns are operated continuously, batch absorption is important in the measurement of reaction rate constants and mass transfer coefficients. These absorption equipment are often carried out in a tank where gas bubbles are dispersed in the liquid phase as shown in Figure 6.1.

6.1 Absorption Dynamics

The approximate differential equation used for calculation of reaction and mass transfer rate is given below[85]

$$\frac{dC_i}{dt} = S_g K_L (C_i^* - C_i) + r_i \qquad (6.1)$$

where S_g is surface area of gas bubbles per volume of liquid, K_L is the mass transfer coefficient from the bubble to the liquid, r_i is the rate of reaction, C_i is concentration of component i in bulk liquid, and C_i^* is equilibrium concentration of i at bubble-liquid interface which for dilute solution can be calculated using Henry's law.

Henry's law

$$C_i^* = p_i/H \qquad (6.2)$$

where p_i is the partial pressure of component i in the gas and H is the Henry's law constant for component i.

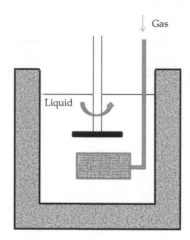

FIGURE 6.1
Batch absorption

6.2 Summary

Absorption involves separation of a solute from gas phase to liquid phase. Absorption can involve reactions also. Batch absorption equipment is used to find reaction rate constants and mass transfer coefficients by measuring concentration changes for the liquid composition with respect to time.

Notations

C_i	concentration of component i [mass/length^3]
C_i*	equilibrium concentration of component i [mass/length^3]
H	Henry's law constant [conc/pressure]
p_i	partial pressure of component i
K_L	mass transfer coefficient
r_i	rate of reaction [mass/length^3/time]
t	integration variable, time [time]
S_g	surface area of gas bubbles per volume of liquid [length $^{-1}$]

7

Batch Extraction

CONTENTS

There are two types of extraction. Solid extraction or leaching involves solids which are leached or extracted by solvent, and liquid extraction where immiscible solvent is used to extract liquid product from the mixture. The following chapter is based on material from [86, 3, 5].

7.1 Solid Extraction or Leaching

There are three processes involved in leaching or solid extraction. Dissolution of solids, separation of solvent from insoluble solid material, and washing. Most of the leaching plants are operated batchwise. In a batch leaching, the solids are stationary and solvent is flowing through the bed of particles.

There are four important factors affecting solid extraction, namely, particle size, solvent, temperature, and agitation. Particle size should be small enough to increase surface area for transfer but large enough to avoid large amounts of fine materials. Since the solubility of material does depend on temperature, it is an important factor to consider. Agitation increases transfer. The solvent should be good selective solvent whose viscosity should be sufficiently low so as to allow circulation.

The commonly used equipment for leaching is the extraction battery which consists of a number of mixer/settler tanks as shown in Figure 7.1. The solids remain in one mixer-settler and the solvent is moved progressively around the ring of tanks. At any time one of the two tanks is out of operation, one being emptied and other being filled. In the remaining tanks ($N - 2$) extraction is proceeding with solvent being passed through the tanks in sequence, the oldest tank receiving fresh liquid and the newest filled with fresh material receiving the most concentrated liquid. After a time interval, the connections are altered so that the tank which is just filled becomes the youngest tank and the former oldest tank out of sequence is being emptied and remaining tanks retain the sequence but with becoming one tank older. In each tank the solvent is referred to as overflow, and a mixture of insoluble residue and solution

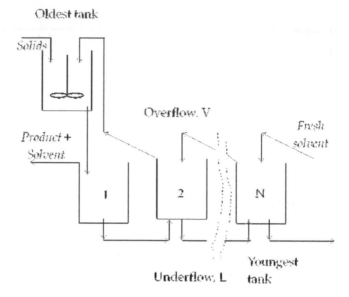

FIGURE 7.1
Schematic of extraction battery

known as underflow is brought into contact with each other. This represents one stage. Figure 7.2 represents N stages of extraction operation. The overflow and underflow leaving the tank are assumed to be in equilibrium and hence have the same composition. This is valid for an ideal stage when the solute is completely dissolved and the concentration of the solution so formed is uniform. Stage efficiency considers factors affecting such a condition. Conventionally, overflow is a lighter stream and underflow heavier. Following the same convention as in distillation, the lighter stream flow is denoted by, V, and composition, y, and the heavier stream flow is represented by L, with composition x. Solution retained by entering solid is then x_0 and solution retained by leaving solid is x_N. Fresh solvent entering the system is y_{N+1} and and concentrated solution leaving the system is y_1. Here, we discuss the ideal stage operation.

Considering overall mass balance around N stages (Figure 7.2) results in:

$$V_{N+1} + L_0 = V_1 + L_N \tag{7.1}$$

The component balance is given by:

$$V_{N+1}y_{N+1} + L_0x_0 = V_1y_1 + L_Nx_N \tag{7.2}$$

Rearranging Equation 7.2 results in:

$$y_{N+1} = \frac{L_N}{V_{N+1}}x_N + \frac{V_1y_1 - L_0x_0}{V_{N+1}} \tag{7.3}$$

where $\frac{L_N}{V_{N+1}}$ represents the slope of operating line. If the viscosity and density of

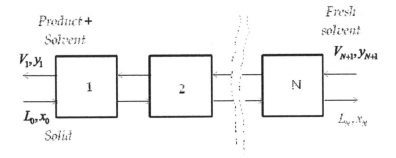

FIGURE 7.2
Schematic of stages in extraction battery

the solution is unaffected by solute concentration, then the following equation for underflow applies.

$$L_N = L_{N-1} = \cdots = L_1 = L \tag{7.4}$$

Constant underflow means constant overflow as given by Equation 7.5 below.

$$V_{N+1} = V_N = \cdots = V_1 = V \tag{7.5}$$

This is the case of constant underflow where slope of the operating line $\frac{L_N}{V_{N+1}}$ is constant for all stages. However, when viscosity and density are functions of solute concentration, the slope changes for each stage and is no longer a linear operating line but a curve. This is the case of variable underflow. In leaching, when sufficient solvent is present to dissolve all the solute in the entering solid, equilibrium is attained when the solute is completely dissolved and solution is uniform. This is the ideal condition and hence the equilibrium relation is given by:

$$y_n = x_n \tag{7.6}$$

Using the equilibrium relation and the operating line given by Equation 7.3, McCabe-Thiele procedure can be performed. This is illustrated in the following example.

Example 7.1: Oil is to be extracted from soyabeans in an extraction battery similar to shown in Figure 7.2 using benzene. If the initial oil content is 15%, and the final extract should contain 40% of oil and 92% of oil is extracted. Calculate the number of stages necessary for this operation. Assume that oil is extracted in the first mixer and equilibrium is reached in each stage. Assume underflow is retained in addition 40% of their weight of solution after each stage.

Solution: We follow the same nomenclature as shown in Figure 7.2 here. Let us assume that 100 kg of raw material (85% beans and 15% oil) is given. The underflow is assumed constant except for the first stage, where the beans are crushed.

Since the oil content in final extract is 92% of original, final oil in extract (in final overflow) is given by:

$$V_1 y_1 = 0.92 \times 15 = 13.8 kg$$

TABLE 7.1

Total mass balance for Example 7.1

Mass in		Mass out	
Underflow	100 kg	*Underflow*	119 kg
Solid	85 kg	Solid	85 kg
Oil	15 kg	Oil	1.2 kg
Solvent	0 kg	Solvent	32.8 kg
Overflow	53.5 kg	*Overflow*	34.5 kg
Solid	0 kg	Solid	0 kg
Oil	0 kg	Oil	13.8 kg
Solvent	53.5 kg	Solvent	20.7 kg

The final extract contains 40% of oil so the final overflow can be given by:

$$y_1 = 0.40$$

$$V_1 = 13.8/0.4 = 34.5kg$$

The solvent in the extract can be calculated as:

$$V_1(1 - y_1) = 20.7kg$$

The final underflow will contain 8% of total oil. Therefore, total oil in final underflow is given by:

$$L_N x_N = 0.08 \times 15 = 1.2kg$$

There are 85 Kg of solid in the underflow which will retain $0.4 * 85 = 34$ kg of solution. Therefore, total underflow is given by:

$$= 85 + 0.4 \times 85 = 119kg$$

Solution in the underflow is given by:

$$L_N = 34kg$$

and

$$x_N = 1.2/34 = 0.035$$

Solvent in the final underflow = 34-1.2= 32.8 kg. Solvent in the extract= 20.7 kg. Therefore, total solvent is given by:

$$V_{N+1} = 53.5kg$$

Total mass balance is shown in Table 7.1.

We start with the first stage:

$$y_1 = 0.4$$

Since the equilibrium line is given by $y_n = x_n$.

$$x_1 = y_1 + 0.4$$

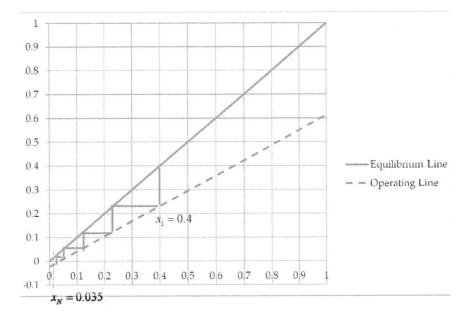

FIGURE 7.3
McCabe-Thiele procedure for Example 7.1

$$L_0 x_0 = 15kg$$

The operating line equation for n-th stage is given by:

$$y_{n+1} = \frac{34}{53.5} x_n + \frac{13.8 - 15}{53.5}$$

Therefore,

$$y_2 = 0.6355 \times 0.4 - 0.0225 = 0.2317$$

Again,

$$x_2 = y_2 = 0.2317$$

Using operating line equation,

$$y_3 = 0.6355 \times 0.2317 - 0.0225 = 0.1247$$

Similarly $x_3 = 0.1247$ and

$$y_4 = 0.6355 \times 0.1247 - 0.0225 = 0.0567$$

As, $y_5 = 0.0135$ and $x_5 = 0.0135 < x_N$, therefore, $N = 5$ stages are required. This is also shown in the McCabe-Thiele procedure in Figure 7.3.

7.2   Rate of Extraction

In extraction, the driving force is the difference between the concentration of the component being transferred, the solute, at the solid interface and in the bulk of the solvent stream. For liquid-liquid extraction, a double film must be considered, at the interface and in the bulk of the other stream.

The rate of change in quantity of solution, dw/dt is given by the following equation

$$\frac{dw}{dt} = K_l A(C_s - C) \tag{7.7}$$

where dw/dt is the rate of solution, K_l is the mass-transfer coefficient, A is the interfacial area, and C_s and C, are the concentrations of the soluble component in the bulk of the liquid and at the interface. It is usually assumed that a saturated solution is formed at the interface and C_s is the concentration of a saturated solution at the temperature of the system.

Fine divisions of the solid component increases the interfacial area A. Good mixing ensures that the local concentration is equal to the mean bulk concentration. In other words, it means that there are no local higher concentrations arising from bad stirring increasing the value of C and so cutting down the rate of solution. An increase in the temperature of the system will, in general, increase rates of solution by not only increasing K_l, which is related to diffusion, but also by increasing the solubility of the solute and so increasing C_s[5].

In the simple case of batch extraction from a solid in a contact stage, a mass balance on the solute gives the equation:

$$dw = V dC \tag{7.8}$$

where V is the quantity of liquid in the liquid stream.

Substituting for dw in Equation 7.7 we then have:

$$V\frac{dC}{dt} = K_l A(C_s - C) \tag{7.9}$$

which can then be integrated over time t during which time the concentration goes from an initial value of C_0 to a concentration y, giving

$$\ln\frac{(C_s - C_0)}{(C_s - C)} = \frac{tK_l A}{V} \tag{7.10}$$

If pure solvent is used then $C_0 = 1$, and:

$$1 - \frac{C}{C_s} = \exp\left(\frac{K_l A}{V}t\right) \tag{7.11}$$

which shows that the solution approaches a saturated condition exponentially. The equation cannot often be applied because of the difficulty of knowing or measuring the interfacial area A. In practice, suitable extraction times are generally arrived at by experimentation under the particular conditions that are anticipated for the plant.

Example 7.2: In a pilot scale test using a vessel $1m^3$ in volume, a solute was leached from an inert solid and the water was 75 percent saturated in 10 seconds.

If, in a full-scale unit, 1000 kg of the inert solid containing, as before, 25 percent by mass of the water-soluble component, is agitated with $100 m^3$ of water, how long will it take for all the solute to dissolve, assuming conditions are equivalent to those in the pilot scale vessel? Water is saturated with the solute at a concentration of 3.0 kg/m^3.

Solution:

For the pilot-scale vessel:

$$C = (3.0 \times 75/100) = 2.25 kg/m^3$$

$$C_s = 3.0 \ kg/m^3, V = 1.0 \ m^3; \ t = 10 \ s$$

Substituting values in Equation 7.11:

$$1 - \frac{2.25}{3.0} = \exp\left(\frac{K_l A}{1.0} 10\right)$$

This results in $K_l A$ is equal to $0.139 m^3/s$.

For the full-scale vessel:

$$C - (1000 \times 25/100)/100 = 2.5 kg/m3$$

Since $C_s = 3.0 kg/m^3$ and $V = 100 m^3$, substituting these values and the value of $K_l A$ in Equation 7.11:

$$1 - \frac{2.5}{3.0} = \exp\left(\frac{0.139}{100} t\right)$$

Therefore,

$$t = 1281.4 \ seconds \ (or \ 21.5 min)$$

7.3 Liquid-Liquid Extraction

Liquid-liquid extraction is complementary to distillation. When distillation is ineffective or very difficult (e.g., azeotropic mixture separation or close boiling mixture separation), or heat requirements in distillation are excessive, extraction is used. In liquid-liquid extraction, two liquid phase separation is used to separate solute from the mixture. The extract is layer of solvent plus extracted solute, and raffinate is the layer from which solute is removed. Most of the extractive equipments are continuous, therefore, we provide a brief description of batch extraction where a mixer-settler tank is used to separate two phases. The two phase composition can be obtained from a ternary diagram. Figure 7.4 shows the ternary diagram for separation of A & B using solvent C. If the circle mark in the figure represents the feed composition then the tie line provides the composition of the extract (represented by a star in the figure) and composition of the raffinate (shown by quadrangle) phases provided enough solvent is added so that the feed composition is in the dome shaped structure in the figure. This represents one equilibrium stage. Sometimes an

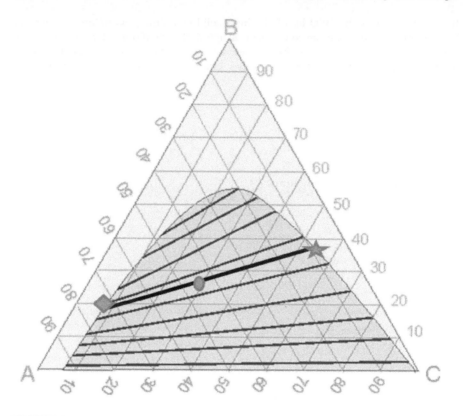

FIGURE 7.4
Ternary diagram and tie lines for liquid liquid extraction (www.et.byu.edu).

extraction battery similar to the one described in the leaching section can be used. Distillation is always preferred compared to extraction because for the recovery of solvent, a distillation column needs to be used after extraction. The more common extraction operation is the extractive distillation. In extractive batch distillation, the solvent is supplied to column continuously or for some time. For details of extractive batch distillation, please refer to the book by Diwekar [7]. Figure 7.5 shows the extractive batch distillation middle vessel column separation sequence for separating industrial solute acetonitrile from water using propyl amine ([1]). In this figure, in the second fraction, entrainer (solvent), E is supplied to the column continuously and in fractions 3 and 4, the entrainer is separated.

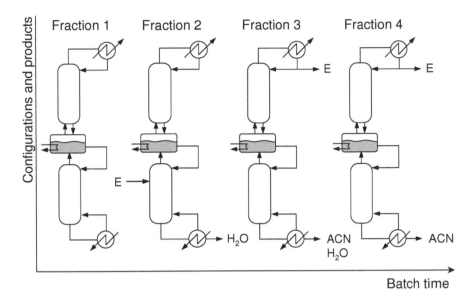

FIGURE 7.5
Operational fractions of batch extractive distillation in a middle vessel column[1].

7.4 Summary

There are two types of extraction processes, namely, leaching where solids are extracted and liquid-liquid extraction. The commonly used leaching equipment is the extraction battery where counter current flow is established resulting in a number of extracting stages. This can be modeled as equilibrium process. The McCabe-Thiele procedure can be used to model this process. In liquid-liquid extraction, ternary diagrams and tie lines are used to determine the extent of extraction. Batch extractive distillation is a commonly used extractive method for liquid-liquid extraction.

Notations

A	interfacial area $[\text{m}^3]$
C	concentration of the soluble component $[\text{mol/length}^3]$
C_s	saturated concentration $[\text{mol/length}^3]$
K_l	mass transfer coefficient $[\text{mass/time}]$
L_j	heavier stream flow leaving stage j $[\text{length}^3/\text{time}]$
N	number of stages

V_j	lighter stream flow leaving stage j [mass/time]
w	quantity of solution [mass]
x_0	composition of solution retained by entering solid
x_j	composition of heavier stream flow leaving stage j
y_j	composition of lighter stream flow leaving stage j
y_{N+1}	composition of fresh solvent entering the system

8

Batch Adsorption

CONTENTS

In batch adsorption a component gets distributed in two phases where one of the phases is solid. The differential affinity of various soluble molecules for specific types of solids is the basis of adsorption. In this process, equilibrium is approached between a solid phase, often called the resin or stationary phase, and the soluble molecules in liquid or gas phase. The liquid or gas phase is then called the mobile phase[6].

8.1 Adsorption Equilibrium

Adsorption equilibrium is a dynamic concept achieved when the rate at which molecules adsorb on to a surface is equal to the rate at which they desorb. The capacity of an adsorbent for a particular adsorbate involves the interaction of three properties: the concentration C of the adsorbate in the fluid phase, the concentration C_s of the adsorbate in the solid phase and the temperature T of the system. If one of these properties is kept constant, the other two may be graphed to represent the equilibrium. The commonest practice is to keep the temperature constant and to plot C against C_s to give an adsorption isotherm. When C_s is kept constant, the plot of C against T is known as an adsorption isostere. In gas solid systems, it is often convenient to express C as a pressure of adsorbate. Keeping the pressure constant and plotting C_s against T gives adsorption isobars[3].

Most earlier theories considered adsorption from gas phase. In principle, the gaseous phase isotherms should be applicable to liquid systems when capillary condensation is neglected. Figure 8.1 show shapes of various isotherms described below.

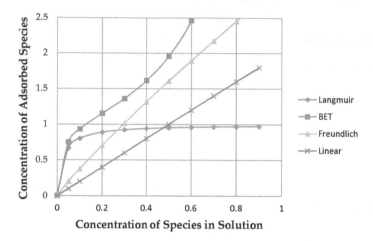

FIGURE 8.1
Shapes of different equilibrium adsorption isotherms

8.1.1   The Linear Isotherm

The simplest isotherm is the linear isotherm. At very low concentrations the molecules adsorbed are widely spaced over the adsorbent surface so that one molecule has no influence on another. For these conditions an isotherm similar to Henry's law for gas liquid systems is given below

$$C_s = K_{eq}C \tag{8.1}$$

where K_{eq} is temperature dependent. Therefore for constant temperature, this is a linear isotherm. However, very few systems will be obeying this isotherm except for very low concentration.

8.1.2   The Langmuir Isotherm

This is the most commonly used isotherm. This adsorption assumes that at higher concentration the rate of adsorption decreases because of lack of space on the adsorbent surface. Therefore, the adsorption is proportional to concentration as well as empty surface available and rate of desorption is proportional to the surface area occupied by adsorbate. The general form of this isotherm is given below

$$C_s = \frac{K_{eq}S_{tot}C}{1 + K_{eq}C} \tag{8.2}$$

where S_{tot} are the total number of sites available.

Figure 8.2 shows a typical Langmuir isotherm. It can be seen from Figure 8.2 that the plot is concave downward, is linear at low concentration, and has a pleateau as the solid is saturated. These kinds of isotherms which are concave downward are called favorable isotherms for adsorption.

This isotherm can also be expressed in terms of partial pressure of adsorbate in the gas phase P or relative pressure P/P^0, where P^0 is the saturated vapor pressure.

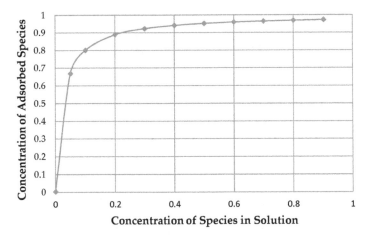

FIGURE 8.2
The Langmuir isotherm

$$C_s = \frac{K_{eq}S_{tot}P}{1 + K_{eq}P} \tag{8.3}$$

If we express the concentration of gas then this equation reduces to Equation 8.4

$$\frac{V}{V'} = \frac{B_1 P/P^0}{1 + B_1 P/P^0} \tag{8.4}$$

where V and V' are equivalent gas phase volume. V' is constant for particular gas.

8.1.3 The Freundlich Isotherm

One of the most popular adsorption isotherm equations used for liquids was proposed by Freundlich in 1926. This isotherm has been used to describe a wide range of of pharmaceutical products like antibiotics, steroids, and hormones [87]. This isotherm is given by the following equation

$$C_s = K_{eq}C^{1/n_f} \quad n_f > 1 \tag{8.5}$$

8.1.4 The BET Isotherm

Brunaur, Emmett, and Teller [88] and Emmett and De witt[89] developed a theory which can account for various shapes of isotherms shown in Figure 8.3. This theory is called BET theory and is based on multi-layered adsorption. It assumes that different layers of adsorptions build up at different parts of the surface by adsorption and desorption (At any time, adsorption creates monolayers on empty surface or from desorption of additional layers). This results in Langmuir isotherm (Type I in Figure 8.3) when just a single layer is used.

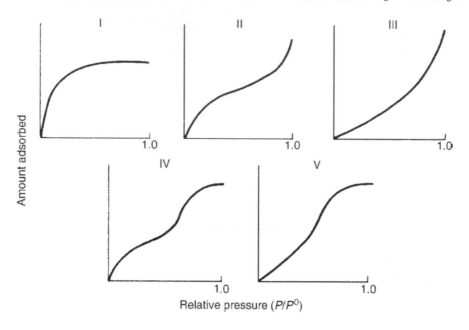

FIGURE 8.3
Five types of BET isotherms[2]

The generalized equation for these five types of isotherms is given by Equation 8.6.

$$\frac{V}{V'} = \frac{B_2 x}{(1-x)} \frac{1 + (1/2ng - n)x^{n-1} - (ng - n + 1)x^n + 1/2ngx^{n+1}}{1 + (B_2 - 1)x + (1/2B_2g - B_2)x^n - 1/2B_2gx^{n+1}} \tag{8.6}$$

where B_2 and g are constants, n represent number of layers and $x = P/P^0$.

For the first three types of isotherms g is assumed to be 2. This results in the following equation. If the adsorption takes place in a limited space, so that even at saturation pressure only a limited number of layers can be adsorbed on the adsorbent, one obtains this expression.

$$\frac{V}{V'} = \frac{B_2 x}{(1-x)} \frac{1 - (n+1)x^n + nx^{n+1}}{1 + (B_2 - 1)x - B_2 x^{n+1}} \tag{8.7}$$

When $n = 1$, this results in Type I or the Langmuir isotherm as stated earlier.

When $n = \infty$ this results in Type II or rarely Type III isotherm given by the Equation 8.8

$$\frac{V}{V'} = \frac{B_2 x}{(1-x)} \frac{1}{1 + (B_2 - 1)x} \tag{8.8}$$

Rearranging the above equation results in Equation 8.9 below.

$$\frac{P/P^0}{V(1 - P/P^0)} = \frac{1}{V'B_2} + \frac{B_2 - 1}{V'B_2} P/P^0 \tag{8.9}$$

Types IV and V suggest that the complete or almost complete filling of the pores and capillaries of the adsorbent occurs at a pressure lower than the vapor pressure of the gas indeed, sometimes at a considerably lower pressure. This lowering of the vapor pressure indicates that as the pressure of the gas is increased some additional forces appear. There is additional heat of evaporation and the term g appears in the equation (Equation 8.6) which is a function of that additional heat. Therefore, for Type IV and V BET isotherms, the term g in Equation 8.6 is temperature dependent. The term B_2 decides the shape. When $B_2 \gg 1$ one obtains Type IV isotherm and for $B_2 < 1$ Type V isotherm.

It has been observed that only gas solid systems provide examples of all the shapes of BET isotherms, and not all occur frequently. For example, charcoal in carbon disulfide, with pores just a few molecules in diameter, almost always gives a Type I isotherm. A non-porous solid is likely to give a Type II isotherm. If the cohesive forces between adsorbate molecules are greater than the adhesive forces between adsorbate and adsorbent, a Type V isotherm is likely to be obtained for a porous adsorbent and a Type III isotherm for a non-porous one.

8.1.5 The Gibbs Isotherm

The Gibbs isotherm assumes that adsorbed layers behave like liquid films, and that the adsorbed molecules are free to move over the surface. This isotherm can be derived then using classical thermodynamics using Gibbs free energy equations. This results in an isotherm of the form given in Equation 8.10. This is known as the Harkins-Jura (HJ) equation[90]. For details of derivation please refer to [3].

$$\ln P/P^0 = L' - \frac{M'}{V^2} \tag{8.10}$$

where L' and M' are constants dependent on surface area of adsorbent.

Example 8.1: The data for volume adsorbed for nitrogen with respect to relative pressure P/P^0 is shown in Figure 8.4. Find out which isotherm fits better for this system.

Solution: The data is for gas solid adsorption so we will concentrate on Langmuir, BET, and Gibbs isotherms for this purpose.

Langmuir isotherm: The equation for the Langmuir isotherm in terms of pressure and volume is given by the following equation

$$\frac{V}{V'} = \frac{B_1 P/P^0}{1 + B_1 P/P^0}$$

This equation can be written in linear form as follows

$$\frac{P/P^0}{V} = \frac{P/P^0}{V'} + \frac{1}{B_2 V'}$$

Figure 8.5 shows the linear regression fit to the data for Langmuir isotherm. It can be seen that the fit is good only for half the points.

For BET isotherm, we will only consider the first three types shown in Figure 8.3. The linear form for this isotherm is given below.

$$\frac{P/P^0}{V(1 - P/P^0)} = \frac{1}{V'B_2} + \frac{B_2 - 1}{V'B_2}P/P^0$$

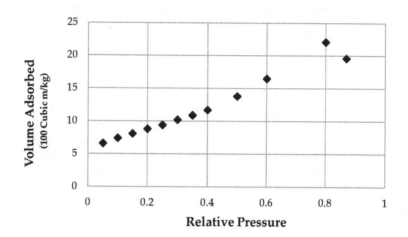

FIGURE 8.4
Data for adsorption of nitrogen

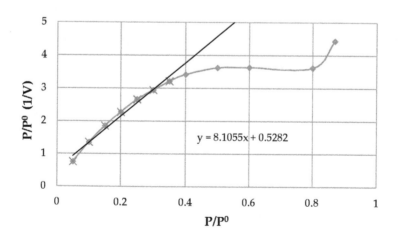

FIGURE 8.5
Langmuir isotherm fit to the adsorption data for nitrogen

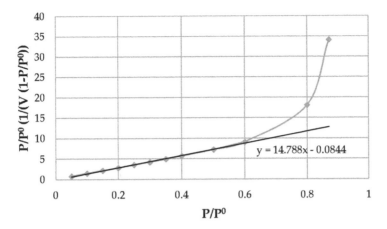

FIGURE 8.6
BET isotherm fit to the adsorption data for nitrogen

Figure 8.6 shows the fit for BET isotherm for the adsorption of nitrogen data. As can be seen, this is a better fit than the Langmuir fit shown earlier.

The Gibbs isotherm is given by:

$$\ln P/P^0 \;=\; L' \;-\; \frac{M'}{V^2}$$

Therefore, the log of relative pressure is plotted against $1/V^2$ to get the fit as shown in Figure 8.7. It can be seen that this isotherm fits almost all points and is the best isotherm to represent this system.

8.1.6 Multicomponent Isotherms

The isotherms presented earlier are applicable to a single component system. The BET isotherm and the Gibbs isotherm are modified for multiple components only for some particular cases and could not be generalized. On the other hand linear, Langmuir, and Freundlich isotherms can be generalized for multicomponent systems and are given below for component i.

Linear Isotherm:

$$C_s \;=\; K_{eq,i}C \tag{8.11}$$

Langmuir[91] Isotherm:

$$\frac{V_i}{V_i'} \;=\; \frac{B_i P_i/P^0}{1 + \sum_i B_j P_j/P^0} \tag{8.12}$$

Freundlich [92] Isotherm:

$$C_{s,i} \;=\; K_{eq,i}C_i^{1/n_{f,i}} \tag{8.13}$$

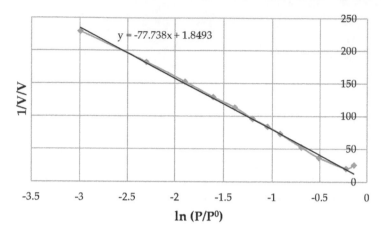

FIGURE 8.7
Gibbs isotherm fit to the adsorption data for nitrogen

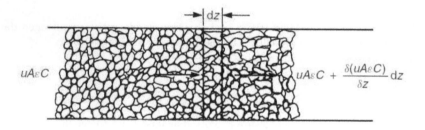

FIGURE 8.8
Cross section of the adsorption column reproduced from [3]

8.2   Fixed Bed or Packed Bed Adsorption

The fixed bed adsorber is the most commonly used arrangement for adsorption and is discussed here. In this vessel the adsorbent is fixed whilst the inlet and outlet positions for process and regenerating streams are moved when the adsorbent is saturated. If continuous operation is required, the unit must consist of at least two beds, one of which is online whilst the other is being regenerated[3]. To understand the dynamics of a fixed bed adsorption column, a mass balance is performed around a disk of cross-sectional area to that of the column (A) but with differential thickness dZ as shown in Figure 8.8.

Mass balance around the cross section is given by:

$$Accumulation \quad = \quad Input - Output - losses$$

$$\frac{\delta(\epsilon u ACdZ)}{\delta t} \quad = \quad uA\epsilon C - [uA\epsilon C + \frac{\delta(uA\epsilon C)}{\delta Z}dZ] + loss \qquad (8.14)$$

where u is the interstitial velocity, that is the velocity of fluid in void fraction ϵ. The rate of loss by adsorption from the fluid phase equals the rate of gain in the adsorbed phase and is given by:

$$\text{Rate of Adsorption} \quad = \quad \frac{\delta\left((1-\epsilon)AC_s dZ\right)}{\delta t} \tag{8.15}$$

Substituting Equation 8.15 in Equation 8.14 and rearranging results in:

$$\frac{\delta(uC)}{\delta Z}|_t + \frac{\delta C}{\delta t}|_Z \quad = \quad -\frac{1-\epsilon}{\epsilon}\frac{\delta C_s}{\delta t}|_Z \tag{8.16}$$

In this equation, the dispersion is completely ignored. If dispersion term is included in this equation then Equation 8.16 will have an additional term as shown in Equation 8.17.

$$\frac{\delta(uC)}{\delta Z}|_t + \frac{\delta C}{\delta t}|_Z \quad = \quad -\frac{1-\epsilon}{\epsilon}\frac{\delta C_s}{\delta t}|_Z + D_{eff}\frac{\delta^2 C}{\delta Z^2} \tag{8.17}$$

where D_{eff} is the effective dispersivity of the separand in the column. For most part, we will be using Equation 8.16 for fixed bed adsorption dynamics.

Since we know that adsorption isotherm can be expressed in terms of function f as $C_s = f(C)$, we can modify Equation 8.16 as follows

$$\frac{\delta uC}{\delta Z}|_t + \frac{\delta C}{\delta t}|_Z \quad = \quad -\frac{1-\epsilon}{\epsilon}\frac{\delta C_s}{\delta C}\frac{\delta C}{\delta t}|_Z$$

$$u\frac{\delta C}{\delta Z}|_t + \frac{\delta C}{\delta t}|_Z \quad = \quad -\frac{1-\epsilon}{\epsilon}f'(C)\frac{\delta C}{\delta t}|_Z \tag{8.18}$$

Rearranging Equation 8.18 results in:

$$\frac{\delta Z}{\delta t}|_C \quad = \quad \frac{u}{1 + \frac{1-\epsilon}{\epsilon}f'(C)} \tag{8.19}$$

This equation shows how the shape of the adsorption wave changes as it moves along the bed. If an isotherm is concave to the fluid concentration axis (e.g., Langmuir isotherm) it is termed favorable, and points of high concentration in the adsorption wave move more rapidly than points of low concentration. Since it is physically impossible for points of high concentration to overtake points of low concentration, the effect is for the adsorption zone to become narrower as it moves along the bed. It is, therefore, termed self-sharpening. An isotherm which is convex to the fluid concentration axis is termed unfavorable. This leads to an adsorption zone which gradually increases in length as it moves through the bed. For the case of a linear isotherm, the zone goes through the bed unchanged. This is shown in Figure 8.9. This equation is also applicable to a non-equilibrium case also.

For cases of non-equilibrium, if a linear driving force for mass transfer is assumed then the following dynamic equations can be written for concentration of separand in stationary phase.

$$\frac{\delta C_s}{\delta t} \quad = \quad K_a(C - C^*) \tag{8.20}$$

$$C_s \quad = \quad f(C^*) \tag{8.21}$$

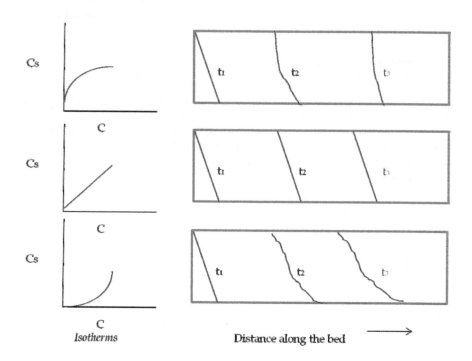

FIGURE 8.9
Shape of adsorption wave for different isotherms

K_a is the mass transfer coefficient and f is the function describing a particular isotherm providing a relation between solute concentration in stationary phase (C_s) and equilibrium composition of solute in mobile phase. Equation 8.16, 8.20, and 8.21 can be solved simultaneously with the following boundary conditions to obtain breakthrough curves (C versus t) for adsorption. This is illustrated in the following example for linear isotherm.

$$t = 0, all\ Z \quad C_s = 0.0 \tag{8.22}$$

$$t > 0, Z = 0 \quad C = C_0 \tag{8.23}$$

C_0 is the initial concentration of solute in the mobile phase.

Example 8.2: A solution containing biological compound was fed at an interstitial velocity of 20 cms/hour to a fixed bed column. The feed composition is 4 milligrams per milliliter, low enough to have a linear adsorption isotherm given below

$$C_s = 4C^*$$

The mass transfer coefficient K_a for this system is approximately equal to 105 h^{-1}. The void fraction of packing is 0.66. Find the breakthrough curves for the column length 0-0.05 cms.

Solution: This is a problem where the dynamics of fixed bed result in partial differential equations given below

$$\frac{\delta C_s}{\delta t} = 105(C - C^*)$$

$$C_s = 4C^*$$

and

$$20\frac{\delta C}{\delta Z}|_t + \frac{dC}{dt}|_Z = -\frac{1 - 0.66}{0.66}\frac{\delta C_s}{\delta t}|_Z$$

$$t = 0, all\ Z \quad C_s = 0.0$$

$$t > 0, Z = 0 \quad C = C_0$$

These partial differential equations can be converted to ordinary differential equations using orthogonal collocation. For this purpose, we will use the shifted Legendre polynomials whose roots and matrices are provided below.

Roots and Matrices of Shifted Legendre Polynomials (Finlayson[21], 1972)

$$N = 1 \quad x = \begin{bmatrix} 0 \\ 0.5000 \\ 1 \end{bmatrix} \quad W = \begin{bmatrix} \frac{1}{6} \\ \frac{2}{3} \\ \frac{1}{6} \end{bmatrix}$$

$$A = \begin{bmatrix} -3 & 4 & -1 \\ -1 & 0 & 1 \\ 1 & -4 & 3 \end{bmatrix} \quad B = \begin{bmatrix} 4 & -8 & 4 \\ 4 & -8 & 4 \\ 4 & -8 & 4 \end{bmatrix}$$

Since this is an illustrative example, let us find the breakthrough curves from t=0 to t=0.5, and Z=0 to Z=0.05 cms. In order to use the above polynomial roots, we need to introduce a new variable $\tau = 2t$ so that we can integrate from $\tau = 0$ to $\tau = 1.0$. This results in following partial differential equations in τ and Z

$$\frac{\delta C_s}{\delta \tau} = 105/2(C - C^*)$$

$$C_s = 4C^*$$

and

$$20/2\frac{\delta C}{\delta Z}\Big|_t + \frac{\delta C}{\delta \tau}\Big|_z = -\frac{1 - 0.66}{0.66}\frac{\delta C_s}{\delta \tau}\Big|_z$$

$$\tau = 0, all\ Z \quad C_s = 0.0$$
$$\tau > 0, Z = 0 \quad C = C_0$$

These equations can be expressed using the following polynomial approximation for $\frac{\delta C}{\delta \tau}$ and $\frac{\delta C_s}{\delta \tau}$

$$\frac{\delta C_j}{\delta \tau} = \sum_{i=1}^{3} A_{ji}C_i \quad j = 1, 2, 3$$

$$\frac{\delta C_{s_j}}{\delta \tau} = \sum_{i=1}^{3} A_{ji}C_{s_i} \quad j = 1, 2, 3$$

Substituting these equations in the partial differential equations given above results in the following ordinary differential equations

$$\sum_{i=1}^{3} A_{ji}C_{s_i} = 105/2(C_j - \frac{C_{s_j}}{4}) \quad j = 1, 2, 3$$

$$\frac{\delta C_j}{\delta Z} = -2/20\left(\sum_{i=1}^{3} A_{ji}C_i + \frac{1 - 0.66}{0.66}105/2(C_j - \frac{C_{s_j}}{4})\right) \quad j = 1, 2, 3$$

The boundary conditions results in the following values at the collocation points

$$C_{s_1} = 0.0\ for\ all\ Z$$
$$C_2|_{z=0} = C_3|_{z=0} = 4$$

We can integrate the differential equations for C_j with respect to Z using numerical integration techniques described in Chapter 2. For simplicity, we will use Euler's method with small value of δZ (0.005). The step by step procedure for this is illustrated below.

1. At Z=0 $C_2 = 4, C_3 = 4$, and $C_{s_1} = 0.0$.
2. We need the initial values of C_1, C_{s_2}, and C_{s_3}. We can find those values using the following algebraic equations using the collocation matrix A .

$$-3C_{s_1} + 4C_{s_2} - 1C_{s_3} = 105/2(C_1 - \frac{C_{s_1}}{4})$$

$$-1C_{s_1} + 0C_{s_2} + 1C_{s_3} = 105/2(C_2 - \frac{C_{s_2}}{4})$$

$$1C_{s_1} - C_{s_2} + 3C_{s_3} = 105/2(C_3 - \frac{C_{s_3}}{4})$$

TABLE 8.1
Breakthrough curves for the example

	Z=0	Z=0.005
t=0	3.2867	3.2536
t=0.25	4.0000	3.9779
t=0.5	4.0000	3.9889

3. Take a step δZ. $Z=Z+0.005$
4. Using Euler's integration:

$$C_1 = C_1 + 0.005 \times 2/20 \times (-3C_1 + 4C_2 - 1C_3 + 26.25(C_1 - \frac{C_{s1}}{4}))$$

$$C_2 = C_2 + 0.005 \times 2/20 \times (-3C_1 + 4C_2 - 1C_3 + 26.25(C_2 - \frac{C_{s2}}{4}))$$

$$C_3 = C_3 + 0.005 \times 2/20 \times (-3C_1 + 4C_2 - 1C_3 + 26.25(C_3 - \frac{C_{s3}}{4}))$$

5. Boundary condition $C_{s1} = 0$
6. We can use the following two algebraic equations to find the value of C_{s2} and C_{s3}.

$$-1C_{s1} + 0C_{s2} + 1C_{s3} = 105/2(C_2 - \frac{C_{s2}}{4})$$

$$1C_{s1} - C_{s2} + 3C_{s3} = 105/2(C_3 - \frac{C_{s3}}{4})$$

7. If Z is not equal to 0.05 go to Step 3, otherwise stop.

The values of breakthrough curve (C vs t) for first two values of Z are shown in Table 8.1.

These equations can also be converted into dimensionless numbers [6] given below.

$$\phi = C/C_0$$
$$\psi = C_s/C_{s0}$$
$$C_{s0} = K_{eq}C_0$$
$$\xi = (K_a Z/u)(1 - \epsilon)$$
$$\theta = K_a/K_{eq}(t - \frac{Z}{u})$$

These dimensionless numbers result in simpler partial differential equations given below. The breakthrough curves in terms of dimensionless numbers are shown in Table 8.2.

$$\frac{\delta\phi}{\delta\xi} + \frac{\delta\psi}{\delta\theta} = 0$$

$$\frac{\delta\psi}{\delta\theta} = \phi - \psi$$

TABLE 8.2

Breakthrough curves for the example in terms of dimensionless numbers

$\xi = 0.0000$		$\xi = 0.8925$	
θ	ϕ	θ	ϕ
0.0000	0.8217	-0.0004	0.8134
1.3125	1.0000	1.3121	0.9945
2.6250	1.0000	2.6246	0.9972

Here we are only showing one integration step because Euler is the simplest numerical integration technique and may not be suitable for these equations for longer horizon.

The other important adsorption unit operation used for bioprocesses is the agitated bed adsorber. In this process, the liquid (e.g., cell culture broth) is passed through a series of agitated columns containing adsorbent. The simple mathematical equation for the $n - th$ column in the series is given below.

$$QC_{n-1} - QC_n = V_L \frac{dC_n}{dt} + V_s \frac{dC_{s_n}}{dt} \qquad (8.24)$$

where Q is the volumetric flowrate coming out of each column, V_L volume of liquid in the column, V_s volume of adsorbent in the column, and C_j concentration of separand leaving $j - th$ column.

8.3   Summary

Adsorption is based on differential affinity of various soluble molecules to solid adsorbent. The equilibrium relation between the solid phase and the liquid or gas phase can be determined using adsorption isotherms. There are various kinds of adsorption isotherm correlations given in the literature, namely, linear, Langmuir, Freundlich, BET, and Gibbs. The most commonly used adsorption isotherm is the Langmuir isotherm. Fixed bed adsorption is commonly used for batch adsorption processes. Dynamics of this process result in partial differential equations result of which shows how the breakthrough curve (concentration profile) is moving through the bed.

Notations

A	cross sectional area [length^2]
B_2	constant in the BET isotherm
C	concentration C of the adsorbate in the fluid phase [mol/length^3]
$C*$	equilibrium composition of adsorbate in the mobile phase [mol/length^3]
C_s	concentration C of the adsorbate in the solid phase [mol/length^3]
g	constant in the BET isotherm
K_a	mass transfer coefficient
K_{eq}	equilibrium constant
L'	constant in the Gibbs isotherm
M'	constant in the Gibbs isotherm
n_f	exponent in Freundlich isotherm
P	partial pressure of adsorbate in the gas phase [mass length^{-1}time^{-2}]
P^0	saturated vapor pressure [mass length^{-1}time^{-2}]
Q	volumetric flowrate [length^3]
t	integration variable, time [time]
S_{tot}	total number of available sites
T	temperature, [^{0}K]
u	interstitial velocity [length/time]
V	gas phase volume at pressure P [length^3]
V_L	volume of liquid in the column [length^3]
V_s	volume of adsorbent in the column [length^3]
V'	gas phase volume at pressure P^0 [length^3]
Z	length of column [length]

Greek Letters:

ϵ	void fraction

9

Batch Chromatography

CONTENTS

Chromatography and adsorption work by the same principle of differential adsorption of species to a surface from a complex chemical mixture. They are known to have high selectivity and can be used for separate components with closely similar physical and chemical properties. Chromatographic techniques can be used to measure a wide variety of thermodynamic, kinetic, and physico-chemical properties. Traditionally, a chromatographic column contains a *stationary phase*, may be a packed bed of solid particles or a liquid with which the packing is impregnated. Then the mixture to be separated is carried in the column dissolved in a gas or liquid stream called the *mobile phase, eluent, or a carrier*. Separation occurs because of the differing distribution coefficients of the components in the mobile and and stationary phases resulting in differing velocities of travel. This is known as *elution chromatography*. This is commonly used chromatographic separation and is described in this chapter. In elution chromatograph, a batch of feed mixture along with the eluent is introduced into the column inlet. The mobile phase causes the band of feed to migrate and split progressively into its component solute-bands or peaks as shown in Figure 9.1[3]. The figure showing these solute-bands is known as a chromatogram (left side of Figure 9.1).

Chromatographic methods are classified according to the nature of the mobile and stationary phases used. The terms gas chromatography (GC) and liquid chromatography (LC) refer to the nature of the mobile phase.

9.1 Retention Theory

Retention theory provides the estimate of time the solute is retained in the column (t_R). A part of this time t_M is required by the solute simply to pass through the mobile phase from inlet to outlet and is measured as the retention time of a non-sorbed solute as shown in Figure 9.2. The adjusted retention time time t'_R represents the extra retention due to repeated partitioning or distribution of the solute between

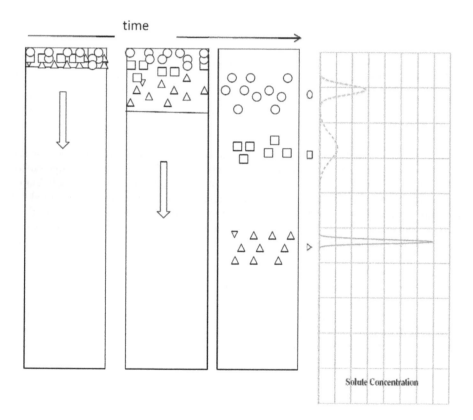

FIGURE 9.1
Schematic illustration of elution chromatography. Three solutes are separating depending on the affinity to stationary phase at different times.

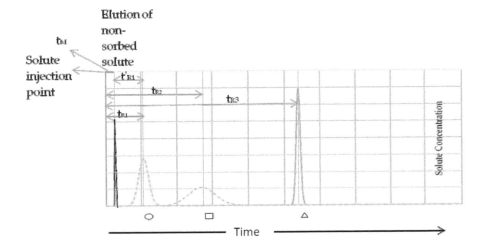

FIGURE 9.2
Chromatogram obtained by elution chromatography of a mixture of three solutes. The retention time t_R is the time taken by a solute to pass through the column. t_M is the mobile-phase holdup and is measured as the retention time of a non-sorbed solute. t'_R is the adjusted retention time, the total time spent by the solute in the stationary phase.

mobile and stationary phases as the band migrates along the column[3]. The ratio R is defined as the fractional time the average molecule spends in the mobile phase is given by:

$$R = \frac{t_M}{t_R} \qquad (9.1)$$

$$= \frac{u_R}{u} \qquad (9.2)$$

where u_R is the velocity at which the solute band moves along the column and u is the velocity of the mobile phase; that is, $u = (superficial\ velocity)/\epsilon$, where superficial velocity is volumetric flow rate divided by cross-sectional area of column and ϵ is the fractional volume of column occupied by mobile phase.

The R can also be defined as the fraction of the total number of molecules that arc in the mobile phase at equilibrium and is given by:

$$R = \frac{n_m}{n_m + n_s} \qquad (9.3)$$

$$= \frac{1}{1 + k'} \qquad (9.4)$$

where n_m is number of molecules in mobile phase and n_s is number of molecules in stationary phase. k' is the mass distribution coefficient. n_s/n_m is usually known as the capacity factor of the column.

Equating Equation 9.4 with Equations 9.1 and 9.2 results in:

$$t_R \quad = \quad t_M(1+k') \tag{9.5}$$

$$u_R \quad = \quad \frac{u}{1+k'} \tag{9.6}$$

Following the same nomenclature as in adsorption (since the fundamental phenomena for chromatography) is the same as adsorption, if C_s is the concentration of solute in stationary phase and C is the concentration in the mobile phase. Then the distribution coefficient between the two phases $K = C_s/C$ can be given by:

$$\frac{n_s}{n_m} \quad = \quad \frac{1-\epsilon}{\epsilon}K \tag{9.7}$$

$$t_R \quad = \quad t_M(1+\frac{1-\epsilon}{\epsilon}K) \tag{9.8}$$

These equations allow us to find the retention times from the distribution coefficient. It has been assumed that the solute follows the linear isotherm. For analytical chromatography, this is a good assumption. For other isotherms, a similar treatment of conservation of mass over a differential length of column as shown in the chapter on adsorption can be used.

9.2 Plate Model

We are familiar with the concept of a theoretical plate or an equilibrium stage. Martin and Synge [93] introduced the concept of number of theoretical plates N in chromatographic separation by dividing the column into a series of imaginary well-mixed tanks at equilibrium and computing the mass balance around each plate. They found out that when the number of plates are large for a solute observing linear adsorption (for low concentrations, this is true as seen in the adsorption chapter), the concentration profile with respect to time can be described by a Gaussian curve as shown in Figure 9.3. It has been found that the number of plates are related to width and retention time given by the following equations (See Figure 9.3).

$$N \quad = \quad \frac{t_R^2}{\sigma^2} \tag{9.9}$$

$$= \quad \frac{t_R^2}{(w/4)^2} \tag{9.10}$$

where w is the width and σ is the standard deviation of the normal distribution.

The HETP is defined as a unit of column length sufficient to bring the solute in the mobile phase issuing from it into equilibrium with that in the stationary phase throughout the unit. Plate models, using this concept, show that the HETP H of a column of length L may be determined by injecting a very small sample of solute, measuring its retention t_R and band width w at the column outlet as shown in

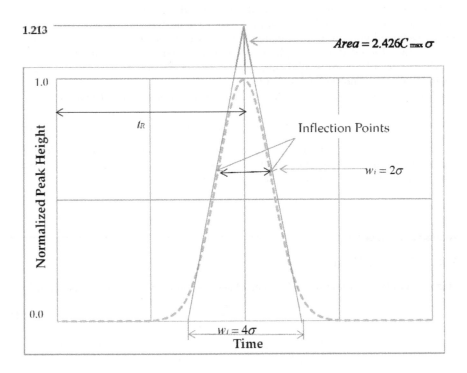

FIGURE 9.3
Properties of Gaussian peak

Figure 9.3, and using the relation:

$$H = \frac{L}{N} \tag{9.11}$$

$$= \frac{Lw^2}{16t_R^2} \tag{9.12}$$

The greater the ratio t_R/w, then the greater the number of theoretical plates N in the column.

To maximize separation efficiency requires low H and high N values. In general terms this requires that the process of repeated partitioning and equilibration of the migrating solute is accomplished rapidly. The mobile and stationary phases must be mutually well dispersed. This is achieved by packing the column with fine, porous particles providing a large surface area between the phases ($0.54\ m^2/g$in GC, 200800 m^2/gin LC). Liquid stationary phases are either coated as a very thin film (0.051 m) on the surface of a porous solid support (GC) or chemically bonded to the support surface as a monomolecular layer (LC)[3].

Van Deemter et al.[94] provided an empirical correlation for H given below.

$$H = A + \frac{B_1}{u} + B_2 u \tag{9.13}$$

In gas chromatography the B terms are usually larger than the A term, while, the A, B_2 terms tend to be dominant in LC. A low concentration of a moderately bound solute should be used to evaluate the column HETP.

9.3   Resolution

Resolution (R_s) is a measure of extent of separation of two peaks and is given by:

$$R_s = \frac{t_{R2} - t_{R1}}{\frac{1}{2}(w_2 + w_1)} \tag{9.14}$$

Thus, resolution can be increased by increasing the difference in retention time and decreasing the peak widths. The resolution Equation 9.14 may be expressed in a more useful form by introducing a separation (selectivity) factor α (similar to relative volatility in distillation):

$$\alpha = \frac{k_2'}{k_1'} \tag{9.15}$$

By substituting values of t_R from Equation 9.5 in Equation 9.15 results in:

$$t_{R2} - t_{R1} = t_M[2k'\frac{\alpha+1}{\alpha-1}] \tag{9.16}$$

Similarly using equations 9.5, 9.10, and 9.14, the equation of resolution R_s results in:

$$R_s = \frac{(\alpha-1)}{2(\alpha+1)}\frac{k'}{1+k'}\sqrt{N} \tag{9.17}$$

This equation provides basic equation for elution chromatography. For better separation, α, N should be large. In chromatographic separation, large N can be readily achieved. A typical large-scale GC or LC column will contain of 103104 theoretical plates.

Equation 9.17 also shows that resolution increases with increasing capacity factor k' but diminishing returns apply at high k' and values of 15 are generally advocated.

9.4 Summary

Chromatography is based on the same principle as adsorption. Chromatographic techniques can be used to measure a wide variety of thermodynamic, kinetic, and physico-chemical properties. Separation occurs because of differing velocities of travel of components in the mobile and stationary phases. Theoretical plate model can be applied to chromatography. Chromatographic methods can be classified as gas and liquid chromatography depending on the nature of the mobile phase.

Notations

A	constant in the HETP equation [length/plate]
B_1	constant in the HETP equation [time/plate]
B_2	constant in the HETP equation [length^2/plate/time]
C	concentration of solute in mobile phase [mass/time]
C_s	concentration of solute in stationary phase [mass/time]
H	HETP [length/plates]
L	length of the column [length]
k'	mass distribution coefficient
K	distribution coefficient between the two phases $K = C_s/C$
n_m	number of molecules in mobile phase
n_s	number of molecules in stationary phase
N	theoretical number of plates
R	fractional time the average molecule spends in the mobile phase
R_s	resolution between two components
t_M	mobile-phase holdup measured as the retention time of a non-sorbed solute [time]
t_R	retention time [time]
t'_R	adjusted retention time [time]
u	velocity of the mobile phase
u_R	velocity at which the solute band moves along the column [length/time]
w	peak width at the base of Gaussian curve[length]

Greek Letters:

α	selectivity factor
ϵ	fractional volume of column occupied by mobile phase.

10

Batch Crystallization

CONTENTS

Crystallization is one of the oldest unit operations where mass is transferred from liquid mixture to solid crystals containing pure solute. Crystallizers are batch, semi-batch, or continuous. Here, we will concentrate on batch crystallization.

Figure 10.1 shows a simple schematic of a batch crystallizer where the still is charged with initial feed and seeded or not seeded. Material crystallizes when the solution is supersaturated. Supersaturation indicates that the concentration of solute is higher than its solubility. Supersaturation can be achieved by four methods. If solubility of the solute increases with temperature as in the case of common inorganic salts and organic substances, then cooling can be used to create supersaturation. When the solubility is not dependent so much on temperature, evaporation is one of the ways for supersaturation (e.g., sugar). If solubility is very high and either cooling or evaporation is not preferable then adding a third component to reduce the solubility of solute can be achieved. The process is known as salting. Lastly, a reactive agent can be used to reduce solubility of the solute. This is called precipitation or reactive crystallization. The most commonly used batch crystallizers are either evaporative or cooling as shown in the schematic in Figure 10.1. Therefore, we will focus this chapter on operation of those crystallizers. The following chapter is based on material from the following literature [95, 86, 3, 5, 96, 97, 98].

10.1 Phase Diagram

A phase diagram provides a feasible region of operation for a crystallizer. Figure 10.2 shows a simple binary phase diagram where temperature and composition are

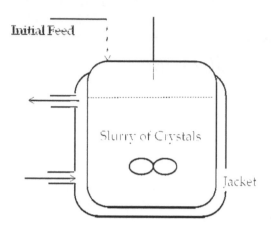

FIGURE 10.1
Schematic of batch crystallizer

plotted on y and x axis, respectively. The point E is the eutectic point and hence the diagram is also called a eutectic diagram. The other important points on the diagram are the freezing temperatures of components A and B. The eutectic diagram indicates what happens when a liquid of a given composition is cooled. At point 1, the mixture of components A and B is in the liquid phase with no solids. At point 2, solid component B starts coming out of solution. As the solution is cooled, more of component B solidifies. As B solidifies, it is removed from the liquid solution. This causes the composition of the liquid to change in such a way that it follows the composition of the liquid. This is shown at point 3 where the overall composition is given by point 3, the solid composition is given by a star, and the liquid composition is given by quadrangle. At the eutectic temperature, the remainder of the material crystallizes at the same composition as the liquid. From a design viewpoint, the eutectic diagram shows that to remove pure component B it is necessary for the initial solution concentration to be between that of the eutectic composition (x_E) and that of pure component ($x_B = 1$). A similar argument can be made for component A where the starting composition must be between that for pure A and the eutectic composition. Other types of eutectic diagrams often exhibit more complex phase behavior; however, the simple eutectic diagram shown in Figure 10.2 represents many systems[95].

10.2   Supersaturation

The basic phenomena influencing crystallization include solid-liquid equilibria and the material will not crystallize unless the solution is supersaturated. Supersaturation is the driving force for the crystallization process and is expressed in terms of concentration. It can be expressed as a difference (ΔC) in the concentration of the

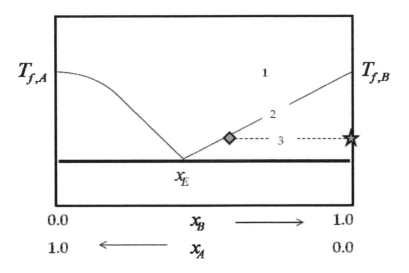

FIGURE 10.2
Eutectic diagram

solute (C) and its saturation concentration (C_s) or as a supersaturation ratio (S); also known as relative supersaturation.

$$\text{Supersaturation} = \Delta C = C - C_s \qquad (10.1)$$

$$\text{Relative Supersaturation} = S = \frac{\Delta C}{C_s} \qquad (10.2)$$

The above thermodynamic information gives an idea about the maximum amount of material that will crystallize as a solid; however, to get an insight into the rate of the production of crystals we need information about its kinetics. The crystallization kinetics provide design information like crystal production rate, size distribution, and its shape.

Figure 10.3 shows the schematic of solubility curve and metastable limit. The metastable zone is shown between the solubility curve and metastable limit. Although supersaturation depends on solubility and supersaturation occurs when ΔC is greater than zero, nuclei may start forming before the supersaturation at any point in the metastable zone. The metastable zone width (MSZW) varies depending on the system being studied. It is usually quite narrow for small ionic crystals, such as NaCl, but can be much wider for organic molecules, such as citric acid.

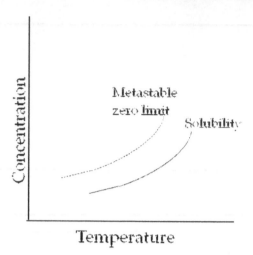

FIGURE 10.3
Schematic of metastable zone

10.3 Crystallization Kinetics

The kinetics are divided into four mechanisms: growth, nucleation, agglomeration, and breakage. Growth refers to the increase in crystal size due to addition of solute molecules to the existing crystals. Nucleation refers to the formation of new solid particles from the solution or formation of small clusters by solute molecules. Agglomeration occurs when two particles collide and stick together to form a larger particle. Breakage occurs in stirred vessels when the larger particle breaks into smaller fragments due to attrition. The phenomena of agglomeration and breakage are rare events and are often neglected while crystallization process modeling.

10.3.1 Growth Rate

Particle growth refers to an increase in particle size due to separation of material from solution. The growth rate is often mass transfer controlled and is given by:

$$G(r,t) \ = \ \frac{dr}{dt} \ = \ \frac{dL}{dt} \tag{10.3}$$

where G is the growth rate and r (or L) is the characteristic dimension for size measurement of a crystal.

The literature on crystallization kinetics [95, 96, 97] have discussed several mathematical expressions depending upon the process complexity for each of the mentioned mechanisms. For example, growth can be size-independent or size-dependent, it can have a constant value, or it may be a function of thermodynamic parameter like solubility. Thus, the selection of proper kinetics specific to the batch crystallization process is a very essential part in process modeling.

One of the first models for crystal growth is McCabe ΔL law and is given by the following size independent rate.

$$G(t) = G_0 \tag{10.4}$$

where G_0 is the parameter which is often a function of temperature and saturation given by the following equation

$$G_0 = K_G(C_i - C_s)^{j_g} \tag{10.5}$$

$$K_G = k_g \exp\left(\frac{-E_g}{RT}/large\right) \tag{10.6}$$

where k_g and E_g are kinetic constants for the equation and R is the gas constant, g is the exponent, and T is the temperature.

10.3.2 Nucleation Rate

Nucleation involves formation of solid particles from solution. Nucleation is characterized by both primary and secondary nucleation rates. The primary nucleation occurs in the solution spontaneously without any solids (seeds) present but secondary nucleation starts in the presence of solid interface. The rates are generally combined as follows.

$$B = B_{primary} + B_{secondary} \tag{10.7}$$

The primary nucleation in general is a function of supersaturation and can be given by the following function

$$B_{primary} = B_{k,1}S^{j_1} \tag{10.8}$$

where $B_{k,1}$ is the rate constant given by $B_{k,1} = A_{N1}\exp(-E_{N1}/RT)$ and j_1 is exponent.

The secondary nucleation is a function of magma density M_T, impeller speed N_I, and length based growth rate G, and and can be written in generalized form as follows

$$B_{secondary} = B_{k,2}N_I{}^h M_T{}^j G^l \tag{10.9}$$

where h, j, l are exponents and $B_{k,2}$ is the rate constant given by $B_{k,2} = A_{N2}\exp(-E_{N2}/RT)$.

10.4 Modeling Cooling Batch Crystallization

Particulate processes are characterized by properties like the particle shape, size, surface area, mass, and product purity. In crystallization the particle size and total number of crystals vary with time. Thus, determining particle size distribution (PSD) is important in crystallization. A population balance formulation describes the process of crystal size distribution with time most effectively. Thus, modeling of a batch crystallizer involves use of population balances to model the crystal size prediction and the mass balance on the system can be modeled as a simple differential equation having concentration as the state variable. The population balance can be expressed as follows[99, 100, 98]

$$\frac{\partial n(r,t)}{\partial t} + \frac{\partial G(t)n(r,t)}{\partial r} = B \tag{10.10}$$

where n is the number density distribution, t is time, r represents characteristic dimension for size measurements, G is the crystal growth rate, and B is the nucleation rate. Both growth and nucleation processes describe crystallization kinetics and their expression may vary depending upon the system under consideration as stated earlier.

Since both growth rate and nucleation rate are concentration dependent, the mass balance in terms of concentration of the solute in the solution is expressed as the following differential equation

$$\frac{dC}{dt} = -3\rho k_v G(t)\mu_2(t) \tag{10.11}$$

where, ρ is density of the crystals, k_v is the volumetric shape factor, and μ_2 is the second moment of particle size distribution (PSD).

Since, $n(r,t)$ represents the population density of the crystals, the i-th moment of PSD is given by Equation 10.12.

$$\mu_i = \int_0^\infty r^i n(r,t)dr \tag{10.12}$$

Each moment signifies a characteristic of the crystal [96]. The zeroth moment corresponds to the particle number, the first moment corresponds to the particle size or shape, the second moment corresponds to its surface area, and the third moment corresponds to the particle volume.

Since population balance equations are multi-dimensional, their calculations are tedious and hence a lot of research has been focused on the model order reduction methods. One of the most common and efficient reduction methods is the method of moments. There are other methods for solving PBEs such as the discretization methods [101], method of characteristics, successive approximation[102], etc. However, the method of moments is commonly used and is described below.

The moment model approach provides a set of ordinary differential equations (ODEs). From the definition of i-th moment in Equation 10.12, we can convert the population balance in Equation 10.10 to moment equations by multiplying both sides by r^i, and integrating over all particle sizes. The moments of order four and higher do not affect those of order three and lower, implying that only the first four moments and concentration can adequately represent the crystallization dynamics[100]. Separate moment equations are used for the seed and nuclei classes of crystals, and are defined as follows

$$\mu_i^n = \int_0^{r_g} r^i n(r,t)dr \tag{10.13}$$

$$\mu_i^s = \int_{r_g}^\infty r^i n(r,t)dr \tag{10.14}$$

where, μ_i is the i-th moment, superscript n, represents nucleation and s represents seed, r_g is the critical radius that distinguishes the two groups. Since we ignore the agglomeration and breakage phenomena, the number of seeds added to the process

(μ_0^s) remain constant and correspond to the initial number of crystals. This facilitates in writing the desired objective function in the optimal control problems discussed at the end of this chapter.

Moment equations for the nucleated crystals [100, 103]:

$$\frac{d\mu_0^n}{dt} = B(t) \tag{10.15}$$

$$\frac{d\mu_i^n}{dt} = iG(t)\mu_{i-1}^n(t), \quad i = 1, 2, \ldots \tag{10.16}$$

Moment equations for seeded crystals [100, 103]:

$$\mu_0^s = constant \tag{10.17}$$

$$\frac{d\mu_i^s}{dt} = iG(t)\mu_{i-1}^s(t), \quad i = 1, 2, \ldots \tag{10.18}$$

The overall i-th moments are defined as summation of nucleated and seeded crystallization moments.

$$\mu_i(t) = \mu_i^n + \mu_i^s, \quad i = 1, 2, \ldots \tag{10.19}$$

Once the moments are available then the data needs to be converted into a number density distribution $(n(r, t))$. The expression given in Equation 10.20 by Flood[104] can be used to calculate the number density from moments

$$\mu_i = \sum_i r^i n_i(r, t) \Delta r \tag{10.20}$$

where $n_i(r, t)$ is the number of particles associated with size bin with r_i and Δr is the size bin. In order to evaluate particle number and size for m number of bins, we need atleast $m + 1$ moments.

Alternatively, one can fit a distribution to the PSD using the moments values [105]. For example, for a normal distribution of a parameter x, the three parameters, namely, mean ($\bar{x}$, coefficient of variation (C_v), and standard deviation (σ) are given by the following expressions. The formula for various distribution can be found in [105].

$$\bar{x} = \frac{\mu_1}{\mu_0} \tag{10.21}$$

$$C_v = \sqrt{\frac{\mu_0 \mu_2}{\mu_1^2} - 1} \tag{10.22}$$

$$\sigma = \bar{x} C_v \tag{10.23}$$

The following example derived from Yenkie et al. [98] illustrates the procedure for modeling and simulation of a cooling batch crystallizer.

Example 10.1: A 100 kg solution containing potassium sulfate and water is to be separated by crystallization where cooling is used to crystallize the potassium sulfate. A linear temperature profile given in Figure 10.4 is used for crystallization. The initial concentration of potassium sulfate in the solution is 0.1743 grams of potassium sulfate/gram of solvent. Find the PSD after 1800 seconds. If initially the crystallizer is seeded with number of crystals equal to 67 and with the crystal

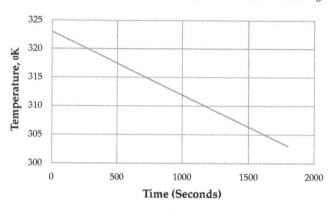

FIGURE 10.4
Linear temperature profile

TABLE 10.1
Parameter values for seeded batch cooling crystallizer

Parameters	Value from experiments / model fitting
	G Kinetics
k_g	1.44 x 108 $\mu m s^{-1}$
E_g/R	4859 ^0K
g	1.5
	B Kinetics
k_g	285 (s m3)-1
E_b/R	7517 ^0K
b	1.45
	parameters related to concentration
ρ	2.66E-12 $g/(\mu m^3)$
k_v	1.5

size distribution given by a normal distribution with mean of 270.96, and standard deviation of 36.785, find how the moments, and concentration are changing with time, and PSD for seeded as well as nucleated crystals at the end of 1800 seconds. Find how much solute is crystallized.

Assume the PSD results in a normal distribution.

This system has been studied by Hu et al.[99], Shi et al.[100], and Yenkie et al.[98], and the data for this system is given below.

Nucleation Kinetics:

$$B(t) = k_b \exp \frac{-E_b}{RT} (\frac{C - C_s}{C_s})^b \mu_3(t) \qquad (10.24)$$

$$G(t) = k_g \exp \frac{-E_g}{RT} (\frac{C - C_s}{C_s})^g \qquad (10.25)$$

The constants in the above equation are given in Table 10.1.

TABLE 10.2
Initial values of moments

Moments	Seed	Nucleation
μ_0	67	0
μ_1	18300	0
μ_2	5050000	0

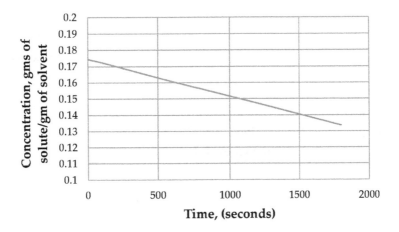

FIGURE 10.5
Concentration profile

The saturation concentration is given by the following equation.

$$C_s(T) = 6.29 \times 10^{-2} + 2.46 \times 10^{-3}T - 7.14 \times 10^{-6}T^2 \qquad (10.26)$$

Solution: From the initial distribution and the number of crystals ($\mu_o^s = 67$) seeded initially, we can calculate the initial values for moments as given in Table 10.2 using Equations 10.15-10.19.

Substituting the initial values from Table 10.2 in Equations 10.15-10.15, along with the equation for concentration (Equation 10.11) and solving saturation concentration equation given by Equation 10.26, using the temperature profile, we can numerically integrate this equations from time $t = 0$ to $t = 1800$ using Runge-Kutta method. Concentration and moments profile for nucleation as well as seeding are given in Figures 10.5-10.8.

The final values of moments are given in Table 10.3.

Using the mean and standard deviation values, the initial and final distributions are shown in Figure 10.9.

The final concentration is found to be 0.134 gms of potassium sulfate/per gram of solvent since the solution is 100 kg with 0.1743 grams/per gram of solvent. Therefore, total solvent S in the mixture is given by:

$$S = 100 \times 1/1.1743 = 85.16kg$$

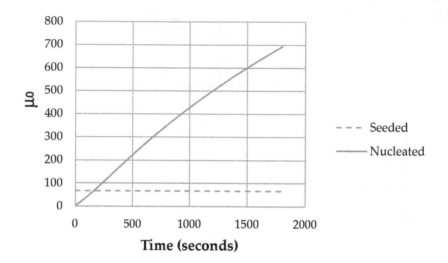

FIGURE 10.6
The zero-th moment profile

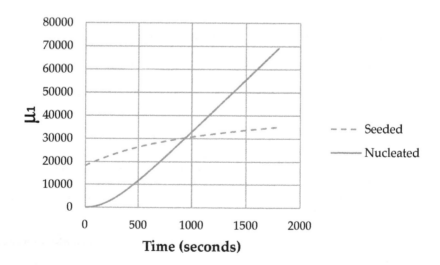

FIGURE 10.7
The first moment profile

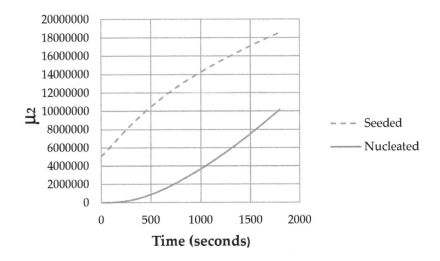

FIGURE 10.8
The second moment profile

TABLE 10.3
Final values of moments

Moments	Seed	Nucleation
μ_0	67	696.1802998
μ_1	135165.1473	69753.1243
μ_2	18580158.05	10300290.28
Mean	527.5299625	100.1940508
σ	21.03288914	68.96801478

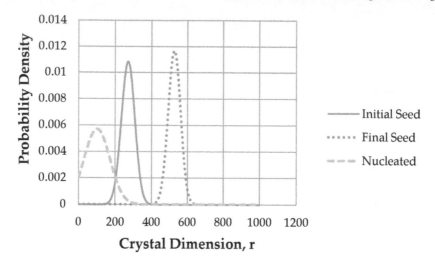

FIGURE 10.9
Particle size distributions

The weight of initial solids in the solution is 100-85.16=14.84 kg. The final concentration is 0.134 grams of solids per gram of solvent. Solids remaining in the final solution are equal to 0.134 × 85.16 = 11.41 kg.

Therefore, potassium sulfate crystallized =14.84-11.41=3.43 kg.

10.5　Modeling Evaporative Batch Crystallization

Modeling evaporative batch crystallization is similar to batch cooling crystallization except one has to consider the rate of evaporation of solvent dM/dt in the equation where M is the total solvent at any time.

The population balance takes the following form

$$\frac{\partial n(r,t)}{\partial t} + \frac{\partial G(t)n(r,t)}{\partial r} + n\frac{d\ln M}{dt} = B \tag{10.27}$$

Moment equations (combining the nucleated and seeded crystals together):

$$\frac{d\mu_0}{dt} = B(t) - \mu_0\frac{d\ln M}{dt} + \mu_0^s \tag{10.28}$$

$$\frac{d\mu_i}{dt} = iG(t)\mu_{i-1}{}^t - \mu_i\frac{d\ln M}{dt}, \quad i = 1, 2, \ldots \tag{10.29}$$

The mass balance of the solute in terms of concentration can be given by:

$$\frac{dC}{dt} = -\frac{\rho k_v}{M}\frac{dM\mu_3}{dt} - C_s\frac{d\ln M}{dt} \tag{10.30}$$

The total heat input rate Q_{in} is given by the following equation

$$Q_in(t) = H_v\frac{dM}{dt} + S_{total}c_{p,soln}\frac{dT}{dt} - H_{cryst}\rho k_v S\frac{d\mu_3}{dt} \qquad (10.31)$$

where H_v is heat of vaporization, H_{cryst} is heat of crystallization, and $c_{p,soln}$ is specific heat capacity of the solution. The heat balance equation is also applicable to the cooling crystallization where the first term is equated to zero.

10.6 Optimal Control Problems

The driving force behind crystallization is the supersaturation of the solute in solution. A critical question in the design and operation of batch crystallization processes is how the supersaturation should vary with time during the batch. This can be achieved by finding optimal temperature profile for cooling crystallizers or by optimal evaporative rate profile for evaporative crystallizers.

Determination of the optimal temperature (or supersaturation) trajectory for a seeded batch crystallizer is a well studied problem. This is a dynamic optimization or optimal control problem. The process performance is determined by the crystal size distribution and product yield at the final time. For uniformity of shape and size in the crystals in a seeded batch crystallization process, it is essential to ensure that the nucleation phenomena occurs to the minimum and mostly the seeded crystals grow to the desired size at a certain rate. If nucleation occurs in the initial phase, then there is a possibility that the nucleated crystal will compete with the seeded ones, thus if the phenomena is of late growth, then nucleation in the earlier phase is preferred. Thus, depending upon the process operation, many types of objective functions have been proposed [4].

Ward et al. [4] divided various objective functions into a list of four objectives as follows:

1. minimize the amount of nucleus-grown crystals,

2. maximize the average size of the total crystals,

3. minimize the variation in size of the total crystals,

4. grow the seed crystals as large as possible.

They argued that many of the supersaturation trajectories found by different authors can be broadly classified into one of the two groups: early growth and late growth. Early growth trajectories are characterized by the fact that supersaturation (and, therefore, growth rate) is greatest at the beginning of the batch, and is reduced toward the end of the batch. Late growth trajectories are characterized by the feature that supersaturation is increasing during the batch and is greatest at the end of the batch. Figure 10.10 is reproduced here from Ward et al. [4] which shows (slightly exaggerated differences) sketches of three different representations of trajectories for both early growth and late growth operating policies.

Since each moment signifies a characteristic of the crystal, depending upon the desired characteristics optimization objectives are simpler to formulate in terms of

FIGURE 10.10
Various trajectories of late and early growth, reproduced from Ward et al. [4]

crystallization moments. Table 10.4 shows these objective functions in terms of moments used by different authors. For each work, Table 10.4 shows the exact mathematical objective function, which word objective(s) [14] the authors were attempting to represent with their mathematical objective, whether they were concerned with low or high moments of the crystal-size distribution, and whether their resulting temperature (or supersaturation) trajectory could be classified as either late growth or early growth.

10.7 Summary

Crystallization is one of the oldest separation processes based on solid liquid phase diagrams. Supersaturation decides the extent of crystallization and is a function of solubility of solute in solution. Supersaturation can be achieved using one of the four methods. Cooling and evaporative crystallization are the two commonly used methods amongst the four methods. In cooling crystallization, temperature is reduced to reach saturation. On the other hand, in evaporative crystallization solvent is evaporated to reduce solubility. Particle size distribution (PSD) is an important factor in crystallization. Method of moments is commonly used to characterize change in PSD with respect to time. The optimal control problems in batch crystallization involve determination of temperature profile for cooling crystallizer and evaporation rate profile for evaporative crystallization. Various objective functions are used for these optimal control problems expressed in terms of moments. Depending on the type of objective function, the profiles generated can be classified as early growth or late growth profiles.

Notations

A_{N1}	primary nucleation rate constant parameter [length/time]
A_{N1}	secondary nucleation rate constant parameter [length/time]
B	nucleation rate [length/time]
$B_{k,1}$	primary nucleation rate constant [length/time]
$B_{k,2}$	secondary nucleation rate constant [length/time]
$B_{primary}$	primary nucleation rate [length/time]
$B_{secondary}$	secondary nucleation rate [length/time]
C	concentration of solute [mol/length^3]
$c_{p,soln}$	specific heat capacity
C_s	saturated concentration [mol/length^3]
C_v	coefficient of variation
ΔC	supersaturation [mol/length^3]
E_g/R	kinetic constant for growth rate [^{0}K]
E_{N1}/R	kinetic constant for primary nucleation rate [^{0}K]

TABLE 10.4

Various objective functions used in optimal control of crystallization[4]

Reference	Mathematical Objective	Word Objective (Min.)	Moments Considered	Operating Policy
[106, 107]	$-\mu_1^s$	4	Low	Late growth
[108]	$-\mu_3$	2	High	Early growth
[108, 109]	$-\mu_1$	2	Low	Early growth
[108]	$\dfrac{-\mu_2}{\mu_0} - \dfrac{-\mu_1}{\mu_0}^2$	2,3	Low	Early growth
[110, 111, 112]	$\dfrac{-\mu_0{}^n}{\mu_1} - \alpha\dfrac{-\mu_1}{\mu_0} \qquad \dfrac{\mu_1{}^s}{\mu_3}$	1,3	High	Late growth
[110, 113]	$\dfrac{-\mu_1}{\mu_4} \qquad \mu_3$	2	Low	Late growth
[110, 98]	$\sqrt{\dfrac{-\mu_2\mu_0}{\mu_1^2} - 1}$	3	Low	Early growth
[114]	$-0.00081\dfrac{\mu_4}{\mu_3} + 0.025\dfrac{\mu_3}{\mu_3} + 0.010\sqrt{\dfrac{-\mu_2\mu_0}{\mu_1^2} - 1}$	1,2,3,4	Both	Late growth
[115]	$\dfrac{\mu_4}{\mu_3} + 0.005\sqrt{\dfrac{-\mu_5\mu_3}{\mu_1^4} - 1}$	2,3	High	Other
[116]	$\dfrac{-\mu_3^n}{\mu_3}$	1	High	Late growth
[117]	$\sqrt{\dfrac{-\mu_5\mu_3}{\mu_4{}^2} - 1}$	3	High	Other

E_{N2}/R	kinetic constant for secondary nucleation rate [^{0}K]
G	growth rate [length/time]
G_0	growth parameter [length/time]
H_v	heat of vaporization
H_{cryst}	heat of crystallization
k_g	kinetic constant [length/time]
K_G	kinetic parameter [length/time]
k_v	the volumetric shape factor
M	total solvent at any time
M_T	magma density
n	number density
N_I	impeller speed [rotations/time]
$Q_{i}n$	total heat input rate [mass length^2 time^{-2}//time]
S	relative supersaturation
r (or L)	characteristic dimension for size measurement
r_g (or L_g)	critical characteristic dimension for size measurement
t	time [time]
T	temperature [^{0}K]
$T_{f,A}$	freezing temperature of A [^{0}K]
$T_{f,B}$	freezing temperature of B [^{0}K]
x_A	composition of A
x_B	composition of B
x_E	eutectic composition
$\bar{x}$	mean value of x

Greek Letters:

μ_i	i-th moment of PSD
μ_i^n	i-th moment of nucleated crystals PSD
μ_i^s	i-th moment of seeded crystals PSD
ρ	density of the crystals
σ	standard deviation

11

Batch Drying

CONTENTS

Drying is often the last separation in a process. Drying refers to removal of water or other solute from solids and often follows evaporation, crystallization, or filtration. Drying is the one of the oldest methods of preservation in the food industry. Drying is often employed to reduce cost of transport, or to have material suitable for handling like in pharmaceutical or soap industries, or in some cases where moisture may cause corrosion.

Drying processes fall into five categories given below.

1. Air or contact drying: In air or contact drying, moisture or other solute is removed from solids by heated air or heated surface.

2. Vacuum drying: In vacuum drying, the fact that evaporation occurs more readily at lower pressure is used for drying.

3. Solvent drying: There are two types of solvent drying processes, namely, superheated-solvent drying and solvent dehydration. In superheated-solvent drying, material containing non-aqueous moisture is dried by contact with superheated vapors of its own associated liquid. This is advantageous when material containing flammable liquids such as butanol is involved [3]. In solvent dehydration water-wet substances are exposed to an atmosphere of saturated organic solvent vapor.

4. Flash drying: In flash drying solid material is in touch with hot gas in a highly turbulent environment for only a short time so that thermal degradation of the product does not occur.

5. Freeze drying: In this process, the material is first frozen and then dried by sublimation (going from solid state to gas state directly without going through the liquid state) in a very high vacuum.

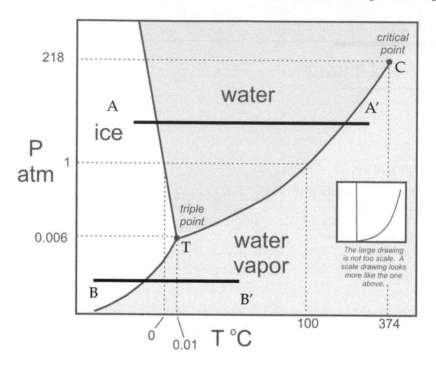

FIGURE 11.1
Phase diagram for water (http://serc.carleton.edu)

11.1   Thermodynamics of Drying

Most of the time, the drying process is used for removing moisture so we will discuss more about thermodynamic properties of water in this section. The phase diagram for water consists of three phases vapor, liquid, and solid as shown in the Figure 11.1. The figure shows three regions of vapor, liquid, and solid separated by each other by various curves. Three phases exist at the triple point, T and at critical point C no phase boundary exists. If heat is applied at any point, the temperature rises and moves horizontally. Below triple point, e.g., point B, the solid phase (ice) will be directly converted to vapor along the line BB'. This is what happens in freeze drying. Above triple point, it has to go to the water phase before going to the vapor phase (as shown in line AA'). Water evaporates when at equilibrium partial pressure of water equals that of vapor pressure at that temperature. The curve TC represents the vapor pressure curve for water is also shown separately in Figure 11.2.

Water content within solids in the two forms, unbound or free water and bound water. Unbound water is in equilibrium with water in phase, i.e., unbound water has the same vapor pressure as bulk water. Unbound water is mainly held in voids of solid. On the other hand, bound water can exist in several conditions: (1) water in fine capillaries that exerts an abnormally low vapor pressure because of the highly

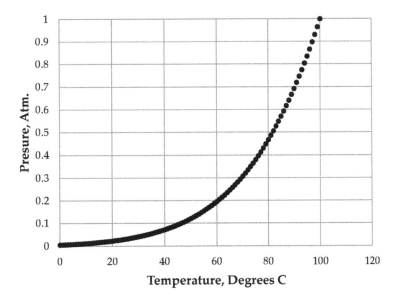

FIGURE 11.2
Vapor pressure temperature diagram for water

concave curvature of surface, (2) water containing a high level of dissolved solids, and (3) water in physical or chemical combination with solids. Solids containing bound water are called hygroscopic[6].

Humidity (absolute humidity) is defined as the mass of water per mass of dry air with units kg kg^{-1}. The capacity of air for moisture removal depends on its humidity and temperature. Equilibrium data of solids in contact with humid air provides information about water capacity of solids. Therefore, typically water content of the solid (insulin here) is plotted as a function of the relative humidity of air as shown in Figure 11.3. Curves of these type are nearly independent of temperature. These curves can be obtained using ideal gas law. From ideal gas law, the concentration of water in vapor form in moles per volume (c_w) can be related to partial pressure of water, p_w and therefore, to humidity as follows

$$c_w = \frac{p_w}{RT} \tag{11.1}$$

$$= \frac{H(M_{air}/M_{water})}{1 + H(M_{air}/M_{water})} \frac{P}{RT}$$

$$= \frac{H(28.9/18.0)}{1 + H(28.9/18.0)} \frac{P}{RT} \tag{11.2}$$

where p_w is the partial pressure, P is the total pressure, T is temperature, R is gas constant, M is molecular weight, and H is humidity. The relative humidity, R_H is the ratio of partial pressure of water to vapor pressure, p_{ws} at that temperature.

$$R_H = \frac{p_w}{p_{ws}} \times 100 \tag{11.3}$$

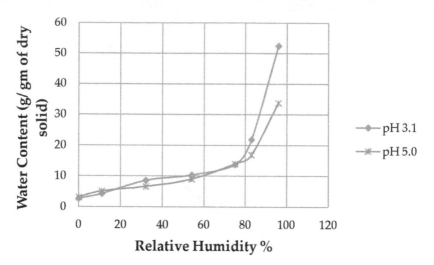

FIGURE 11.3
Equilibrium-moisture curves

Latent heat of vaporization is the energy required to evaporate per kg of liquid (in this case water) and latent heat of sublimation is the energy required for 1 kg of solid to evaporate. These values can be obtained from the steam tables. In carrying out heat calculations for dryer, the enthalpy of drying can be calculated using the following equation

$$E_{dry} \quad = \quad c_p(T - T_0) + H\lambda_0 \qquad (11.4)$$

where E_{dry} is the heat required for drying, c_p specific heat of liquid (solid for sublimation), and λ_0 is the latent heat of vaporization (sublimation for freeze drying) at temperature T_0.

The study of the relationship between air and water is called psychrometry. Psychrometric charts show properties of mixture of air and water as shown in Figure 11.4. Any point on this chart represents specific mixture of water and air. The curved line represents saturation curve for 100 % humidity of air saturated with water as a function of dry bulb temperature.

The dry bulb temperature (DBT) is the temperature of air measured by a thermometer freely exposed to the air but shielded from radiation and moisture. The wet-bulb temperature (θ_w) is the temperature a parcel of air would have if it were cooled to saturation (100% relative humidity) by the evaporation of water into it, with the latent heat being supplied by the parcel. In other words, wet bulb temperature is the temperature reached by water surface if the air is passed over it. Wet bulb temperature is a function of dry bulb temperature and humidity. The chart shows dry bulb temperature on the x-axis and moisture content on the y-axis. Any point below the saturation line represents air that is unsaturated, therefore, the chart has relative humidity curves going up to 100% relative humidity. Wet bulb temperature lines are constant enthalpy or adiabatic cooling lines. The change in composition of

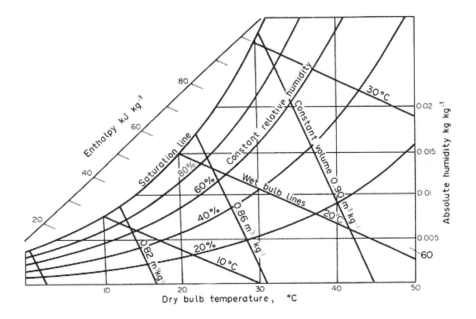

FIGURE 11.4
Psychrometric chart[5]

the mixture when temperature is lowered could be followed by moving along these lines towards 100% saturation curve.

At wet bulb temperature, the heat transfer to air can be given by:

$$Q = hA(\theta - \theta_w) \tag{11.5}$$

where h is the heat trasfer coefficient, A is the surface area, θ is temperature of air, and θ_w is wet bulb temperature.

Using mass transfer rate, the heat transfer at equlibrium can be written as:

$$Q = K_G A \rho_A \lambda (H_w - H) \tag{11.6}$$

where K_G is the mass transfer coefficient, ρ_A is the density of air, H_w is humidity of saturated air at wet bulb temperature, and H is the humidity of gas stream.

Equating equations 11.5 to 11.6 results in the following equation.

$$H - H_w = \frac{h}{h_D K_G \rho_A \lambda}(\theta - \theta_w) \tag{11.7}$$

Equation 11.7 shows how wet bulb temperature is dependent on humidity and temperature.

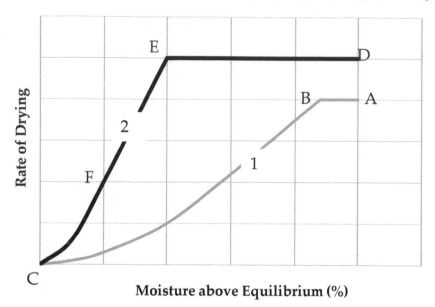

FIGURE 11.5
Drying curve for different materials

11.2   Rate of Drying

In drying, at first the unbound moisture is removed from the surface and then moisture from the interior of the material. A drying curve is a curve representing rate of drying versus moisture content as shown in Figure 11.5. The type of drying curve depends on the structure and type of material. Figure 11.5 shows two typical curves. Curve 1 (e.g., drying curve for soap) shows two well-defined regions: region AB, where rate of drying is constant and BC, where there is a steady fall in the rate of drying. Curve 2 (e.g., drying curve for sand) shows three stages: DE the constant-rate period, EF and FC are falling-rate periods. EF is known as the first falling-rate period and FC the second falling-rate period. The moisture content at the end of constant-rate period (B in curve 1 and E in curve 2) is known as critical moisture content.

Constant-rate Period

.

During constant-rate period, the free moisture leaves the saturated surface of the material by diffusion through a stationary air film into air stream. Gilliland[3] showed that the drying rate for this period for variety of material is essentially the same (around 2.0-2.8 $kg/m^2/h$).

The mass rate of evaporation (W) is found to be dependent on partial pressure

p_w, surface area A, velocity of air stream u, and ratio of length L to width B of the surface and is given below[3].

- For values of $u = 1 - 3 \ m/s$:

$$W \quad = \quad 5.53 \times 10^{-9} L^{0.77} B (p_{ws} - p_w)(1 + 61u^{0.85}) \ [kg/s] \qquad (11.8)$$

- For values of $u < 1 \ m/s$:

$$W \quad = \quad 3.72 \times 10^{-9} L^{0.73} B^{0.8} (p_{ws} - p_w)(1 + 61u^{0.85}) \ [kg/s] \qquad (11.9)$$

However, for most practical purposes drying at a constant-rate period can be assumed to be proportional to $(p_{ws} - p_w)$ given by,

$$W \quad = \quad \frac{dw}{dt} = K_G A(p_{ws} - p_w) \qquad (11.10)$$

where K_G is the mass transfer coefficient.

The simple expression of time required for drying period can be expressed in terms of moisture content w_i and final moisture content equal to critical moisture content w_c as:

$$t_c \quad = \quad \frac{1}{W}(w_i - w_c) \qquad (11.11)$$

where t_c time required for drying during a constant-rate period.

First Falling-rate Period

This period starts at critical moisture content (w_c) below which the surface is no longer capable of supplying sufficient free moisture to saturate the air in contact with it. During this period, the rate of drying is approximately proportional to the free moisture content $(w - w_e)$, where w is the moisture content and w_e is the equilibrium moisture content.

$$-\frac{1}{A}\frac{dw}{dt} \quad = \quad m(w - w_e) \qquad (11.12)$$

Simplifying and integrating:

$$-\frac{1}{mA} \int \frac{dw}{(w - w_e)} \quad = \quad \int dt \qquad (11.13)$$

$$t_f \quad = \quad \frac{1}{mA} \ln \frac{(w_c - w_e)}{(w - w_e)} \qquad (11.14)$$

where t_f time required for drying during falling-rate period. The rate of drying at the end of the constant drying period and the first falling-rate period initial drying rate are the same. Therefore, the time required for the constant-rate period is given by,

$$t_c \quad = \quad \frac{1}{w_c - w_e}(w_i - w_c) \qquad (11.15)$$

Second Falling-rate Period

At the conclusion of the first falling-rate period it may be assumed that the surface is dry and that the plane of separation has moved into the solid. In this case, evaporation takes place from within the solid and the vapor reaches the surface by molecular diffusion through the material. The forces controlling the vapor diffusion determine the final rate of drying, and these are largely independent of the conditions outside the material[3].

11.2.1   Movement of Water during Drying

It is important to understand how the moisture moves to the drying surface during the falling-rate period, and two models have been used to describe the physical nature of this process, the diffusion theory and the capillary theory. In the diffusion theory, the rate of movement of water to the air interface is governed by rate equations similar to those for heat transfer, whilst in the capillary theory the forces controlling the movement of water are capillary in origin, arising from the minute pore spaces between the individual particles.

Diffusion Theory

In the falling-rate period the surface is no longer saturated and hence the rate of drying falls steadily. In many cases, drying is governed by internal movement of water. It was proposed that this can be explained using diffusion theory and can be explained using Fick's second law of diffusion given below in terms of moisture content of solid[3].

$$\frac{dc_w}{dt} = D\frac{\partial^2 c_w}{\partial y^2} \tag{11.16}$$

where c_w is concentration of water in solid expressed as mass water per unit mass of dry solid, D is the diffusion coefficient of water in solid matrix, and y is the distance measured in the direction of diffusion.

Since diffusion theory is applicable to non-porous solids, the following assumptions are made to solve the equation.

- liquid diffusion is independent of moisture content, i.e., D is assumed to be constant.

- initial moisture distribution is uniform.

- material size, shape, and density are constant.

For the falling-rate period the initial moisture content is equal to the critical moisture content and at the surface the moisture content is equal to equilibrium moisture content. The following boundary conditions result for a solid slab with a thickness equal to $2l$

$$at\ t = 0\ and\ -l < y < l, \quad w = w_c \tag{11.17}$$
$$at\ t > 0\ and\ y = \pm l, \quad w = w_e \tag{11.18}$$

Using these boundary conditions, and solving Equation 11.16 results in:

$$\frac{w - w_e}{w_c - w_e} = \frac{8}{\pi^2}\left[\sum_{n=0}^{n=\infty} \frac{1}{(2n+1)^2} \exp\left(-(2n+1)^2 Dt(\frac{\pi}{2l})^2\right)\right] \quad (11.19)$$

For long drying times, when Dt/l^2 exceeds approximately 0.1, only the first term can be considered in the above equation. Differentiating this term results in the following drying rate.

$$-\frac{dw}{dt} = \frac{\pi^2 D}{4l^2}(w - w_e) \quad (11.20)$$

It can be seen that the two equations 11.12 and 11.20 are similar.

Capillary Theory

For granular or porous solids moisture tends to flow by the action of capillary forces rather than diffusion. Capillary forces are higher for higher curvature of the meniscus of the water in the pore. The curvature increases as pore size decreases. Therefore, the small pore sizes have greater capillary forces. The larger pores empty first during the constant-falling-rate period, and the small pores are opened later during the falling-rate period [6]. Opening of pores depends on the suction potential. Equations for suction potential can be obtained from [3].

11.3 Types of Dryers

Dryers can be continuous or batch operated. Batch dryers are less capital cost intensive and flexible. Major categories of dryers fall under four categories: tray or shelf dryers, batch vacuum rotary dryers, freeze dryers, and spray dryers. The following paragraphs describe these different dryers, design and operation.

Tray or Shelf Dryers

Tray or shelf dryers are used for granular material or for individual articles. The material is placed on a series of trays or shelves and heated by steam coils (or circulating warm water) from below or passed heated air over the material. The operation of multi-pass tray or shelf dryer heated by air is shown in Figure 11.6 using the psychrometric chart presented earlier. In this figure, A is the initial point for air with humidity H_1 which is passed over the material and its temperature rises to θ_1 (point B). The air leaves at 90% relative humidity (point C) with its temperature falling to θ_2 removing $H_2 - H_1$ kg of water per kg of air. This operation (line BC) is adiabatic. The air is again reheated to temperature θ_1 (point D) and again passed through the second tray or shelf where the air reaches 90% relative humidity and its temperature rises to θ_3. At this stage air has picked up $H_3 - H_1$ kg of water per kg of air. The figure shows 4 passes reducing the water content by $H_4 - H_1$ kg of water per kg of air. This multi-pass method is beneficial in two ways, first it needs less amounts of air to remove the same amount of water and secondly in order to

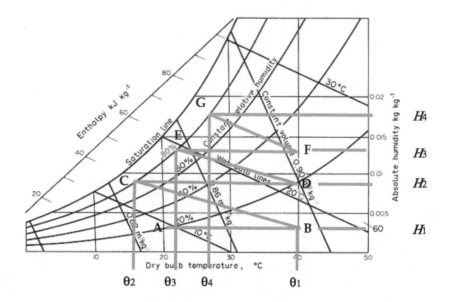

FIGURE 11.6
Multi-pass tray dryer operation heated by air

remove the same amount of moisture in single-pass the air needs to be heated at very high temperature increasing the heat requirements.

For the systems where the heat is conducted through the shelves to the solids, the following analysis could be used for design and operation of a dryer. For heat conduction drying Fourier's law of heat conduction is applicable.

$$q \;=\; -k\frac{dT}{dy} \tag{11.21}$$

where q is the heat flux (flow rate of heat per unit area), T is the temperature, k is the thermal conductivity of solid, and y is measured in the direction of heat flow.

Figure 11.7 shows the schematic of conductive drying in a tray where the shelves are maintained at temperature T_0, and the density of solid is ρ_0. The figure shows receding boundary between the wet and dry solids using a dashed line. Temperature at this stage is T_s the length at which the boundary is at y. l is the width of solid slab. Then the Fourier's law given by Equation 11.21 can be expressed as follows.

$$q \;=\; -k\frac{T_0 - T_s}{y} \tag{11.22}$$

The heat required to vaporize the water is equal to heat conducted resulting in the following equation.

$$qA \;=\; -\lambda\frac{d(\rho_0 y A)}{dt} \tag{11.23}$$

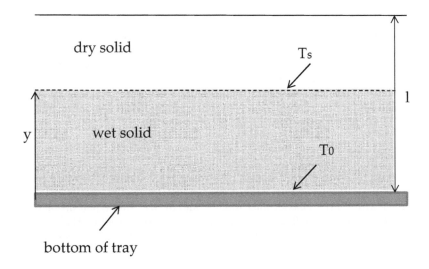

FIGURE 11.7
Schematic of conductive drying in a tray

We know that at $t = 0$ $y = l$ then integrating Equation 11.23 by substituting value of q from Equation 11.22 results in

$$\frac{l^2 - y2}{2} = \frac{k(T_0 - T_s)}{\lambda \rho_0} \tag{11.24}$$

The drying is completed when $y = 0$, which gives the time for drying as

$$t = \frac{l^2 \lambda \rho_0}{2k(T_0 - T_s)} \tag{11.25}$$

Vacuum Rotary Dryers

Another type of dryer where heat conduction is used for drying is vacuum rotary dryers also called a vacuum tumble dryer. Heat is supplied by warm water or any other heat exchange medium circulating through a jacket of rotating-cone drum. Vacuum is applied to be able to dry at lower temperatures and more rapidly.

For scale up of these dryers, small scale dryers are used to obtain the time for drying and the following scale up equation is used.

$$\frac{t_{large}}{t_{small}} = \frac{(A/V)_{small}}{(A/V)_{large}} \tag{11.26}$$

where *small* refers to small dryer and *large* refers to large dryer. A and V are area to volume ratio of the dryer.

Freeze Dryers

There are three steps involved in freeze drying, freezing, primary drying, and secondary drying.

Special considerations apply to the movement of moisture in freeze drying. Since the water is frozen, liquid flow under capillary action is impossible, and movement must be by vapor diffusion, analogous to the second falling-rate period of the normal case. In addition, at very low pressures the mean free path of the water molecules may be comparable with the pore size of the material [3].

Spray Dryers

A spray dryer transforms a feed in the liquid state into a dried particulate form by spraying the liquid into hot gas. Three basic unit processes are involved in spray drying: liquid atomization, gas-droplet mixing, and drying from liquid droplets. The rapidity of drying and the relatively low temperature experienced by the solids during drying have led to the successful application of spray drying for many types of heat sensitive products including milk, coffee, blood, spores, and various microorganisms, and antibiotics [6].

Spray dryers are sized based on residence time of the air in drying chamber. This is a function of chamber volume and total air flowrate. The ratio of height to diameter of the chamber depends on droplet size. The heuristics for minimum chamber diameter is as follows[6].

- 2 m for 40-80 μm drops

- 4 m for 80-100 μm drops

- 5 m for 100-120 μm drops

Theoretical correlations for design are developed by Gluckert [6]. For spray dryers operating with pressure nozzle atomizers, for example, the chamber volume V of the large scale unit can be calculated from the following correlation:

$$Q = \frac{10.98 k_f V^{2/3}(T - T_s)_{out}}{D_m^2} D_s \sqrt{\frac{\rho_1}{\rho_m}} \qquad (11.27)$$

where Q is the rate of heat transfer to spray (BTU/h), k_f is the thermal conductivity of gas film surrounding the drop (BTU/h ft ^{0}F), V is volume of drying chamber (ft^3), T is the gas phase bulk temperature (^{0}F), T_s is temperature at drop surface, D_m is maximum drop diameter, assumed to be three times surface per unit volume average size (ft), D_s diameter of pressure nozzle discharge orifice (ft), ρ_1 is density of gas at exit conditions (lb/ft^3), and ρ_m is density of spray jet at the atomizer (lb/ft^3).

11.4   Summary

Drying involves removal of moisture or other solute from solids. It is often the last separation process. Relation between air and water can be explained using psychrometric charts. The rate of drying of material involves a constant-rate period and one or two falling-rate periods. Diffusion theory and capillary theory are two theories designed to explain the falling-rate period. Main types of dryers can be classified into four categories: tray or shelf dryers, batch vacuum rotary dryers, freeze dryers, and spray dryers.

Notations

A	qrea [length^2]
B	width [length]
c_p	specific heat capacity [length^2 time^{-2}/^{0}K]
c_w	concentration of water in vapor form [mol/length^3]
D	diffusion coefficient of water in solid matrix
D_m	maximum drop diameter [ft]
D_s	diameter of pressure drop [ft]
E_{dry}	enthalpy of drying
h	heat transfer coefficient
H	humidity (absolute humidity)[kg kg^{-1}]
H_w	humidity of saturated air[kg kg^{-1}]
k	thermal conductivity
k_f	thermal conductivity of gas film surrounding the drop [BTU/h ft ^{0}F]
K_G	mass transfer coefficient
l, L	length [length]
m	slope of falling-rate period [length^{-2}]
p_w	partial pressure of water [mass length^{-1}time^{-2}]
p_{ws}	vapor pressure of water [mass length^{-1}time^{-2}]
P	total pressure [mass length^{-1}time^{-2}]
q	heat flux
Q	rate of heat transfer
Q	rate of heat transfer spray [BTU/h]
R	gas constant, 1.98719 [cal/mol/^{0}K]
R_H	relative humidity
t	integration variable, time [time]
t_c	constant-rate period time [time]
t_f	first falling-rate period time [time]
T	temperature [^{0}K]
T_s	temperature at a stage [^{0}K]
T_s	for spray dryer temperature at drop surface [^{0}F]
u	velocity of air stream [length/time]
V	volume of drying chamber [ft^3]
w	moisture content [mass]
w_c	critical moisture content [mass]
w_e	equilibrium moisture content [mass]
W	mass rate of evaporation [mass/time]
y	distance in the direction of heat flux [length]

Greek Letters:

θ	temperature of air [^{0}K]
θ_w	wet bulb temperature [^{0}K]
λ	latent heat of vaporization (or sublimation)
λ_0	latent heat of vaporization (or sublimation) at temperature T_0
θ	temperature of air [^{0}K]
θ_w	wet bulb temperature [^{0}K]
ρ_1	density of gas at exit conditions [lb/ft^3]
ρ_A	density of air [mass/vol]
ρ_m	density of spray jet at the atomizer [lb/ft^3]

12

Batch Filtration

CONTENTS

Filtration involves separation of solids from liquid using a porous medium or screen. Filtration is a mechanical separation and requires less amounts of energy than drying. There are two broad categories of filtration: conventional or dead-end filtration and crossflow or tangential flow filtration. The direction of the fluid feed in relation to the filter medium differentiates between these two categories. When the fluid feed is flowing perpendicular to the medium it is the conventional filtration and when it is parallel to the medium, it is crossflow filtration. In figure 12.1 a schematic of these two categories are shown. In case of conventional filtration, the cake thickness builds up fast as compared to the crossflow filtration (Figure 12.1). The medium used for conventional filtration are filter clothes, meshes, or screens. Crossflow filtration uses membranes for separation.

12.1 Conventional or Dead-End Filtration

The general theory of conventional filtration proposed by Suttle[118] is presented here. In conventional filtration, if the pressure is kept constant then the rate of flow progressively diminishes, whereas if the flowrate is kept constant then the pressure must be gradually increased. Because the particles forming the cake are small and the flow through the bed is slow, streamline conditions are almost invariably obtained, and, at any instant, the flowrate of the filtrate may be represented by Darcy's law given below:

$$\frac{1}{A}\frac{dV}{dt} = \frac{\triangle P}{\mu R} \tag{12.1}$$

where A is the cross sectional area of the exposed filter medium, V is volume of filtrate, t is time, $\triangle P$ is the pressure drop through the bed of solids (medium plus

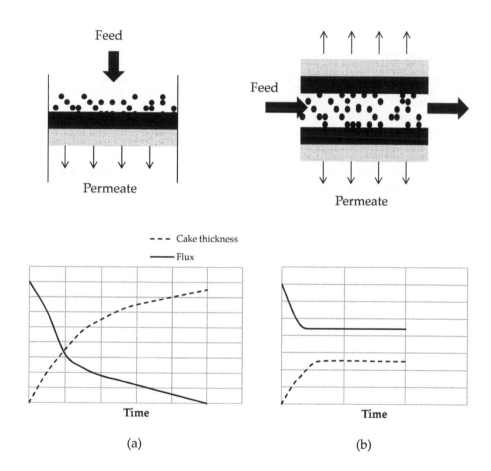

FIGURE 12.1
Schematic diagram of (a) conventional and (b) crossflow filtration operation

cake), μ is the viscosity of filtrate, and R is the resistance of the porous bed. The resistance R consists of resistance of medium, R_m and resistance of cake, R_c.

$$R = R_m + R_c \tag{12.2}$$

The resistance of cake is a function of voidage, e, the specific area of particle, S and the cake thickness, l.

$$R_c = \frac{5(1-e^2)S^2}{e^3}l \tag{12.3}$$

For incompressible fluids the e and S are constant resulting in the following equation

$$R_c = r\rho_c l \tag{12.4}$$

where r is called the specific resistance. For compressible fluids this resistance is a function of pressure drop across the cake, $\triangle P_c$ and is given by

$$r = r'(\triangle P_c)^s \tag{12.5}$$

where r' and s are empirical constants where the power s is called cake compressibility factor. This factor ranges from zero for incompressible cakes such as sand to nearly unity.

Equation 12.1 can be written in terms volume of the filtrate as follows.

$$\frac{1}{A}\frac{dV}{dt} = \frac{\triangle P}{\mu(r\rho_c(V/A)+R_m)} \tag{12.6}$$

In the early stages the rate of filtration is constant as the resistance of the medium much larger than the cake resistance. Integrating Equation 12.1 results in an equation for pressure drop.

$$\triangle P = = \frac{V}{tA}\mu(r\rho_c(V/A)+R_m) \tag{12.7}$$

If the constant rate filtration occurs from $t=0$ to $t=t_1$ then Equation 12.1 can be integrated to get,

$$\frac{V_1}{t_1} = \frac{A\triangle P}{\mu(r\rho_c(V_1/A)+R_m)} \tag{12.8}$$

Once the initial cake is build up, most of the conventional filtration operation use constant pressure drop period for filtration. In this case, the filtration rate falls. Again integrating Equation 12.6 from $t=t_1$ to $t=t$ results in,

$$\frac{t-t_1}{V-V_1} = \frac{\mu r\rho_c(V-V_1)}{2A^2\triangle P} + \frac{\mu r\rho_c(V_1)}{2A^2\triangle P} + \mu\frac{R_m}{A\triangle P} \tag{12.9}$$

TABLE 12.1

Data for filtration

Time(s)	V/A(cms)
29	1.9290
106	3.8580
217	75.7870
376	7.7161
591	9.6451
780	11.5741
1088	13.310

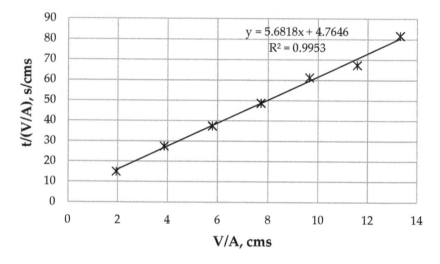

FIGURE 12.2

Filtration curve for Example 12.1

It is clear from the above equation that there is a linear relationship between $\frac{t-t_1}{V-V_1}$ and $V - V_1$ and it does not go through the origin.

Example 12.1: A batch conventional filtration equipment is used to filter a cell culture suspension which has a viscosity of 3.0 cp. The data for fitration (Table 12.1) at constant pressure is obtained with a vacuum pressure of 500 mm HG. The cell solids on the filter at the end of filtration were dried and found to weigh 14 grams. Determine the specific cake resistance, r and the medium resistance R_m. Estimate how long it would take to obtain 5000 liters of filtrate from this cell broth on a filter with a surface area of 12 m^2 and vacuum pressure of 450 mm Hg. The final volume of the cake is found to be 690 ml.

Solution: Data from Table 12.1 is used to generate the $\frac{t}{V/A}$ and V/A graph for this problem. This graph is shown in Figure 12.2. The slope of the line will give us the value of specific cake resistance, r and the intercept can be used to evaluate, R_m.

If V_1 is assumed to be zero, then Equation 12.9 results in Equation 12.10. This equation can be used to find the values of cake specific resistance and media resistance from

$$\frac{t}{V/A} = \frac{\mu r \rho_c (V/A)}{2 \Delta P} + \frac{R_m}{\Delta P} \tag{12.10}$$

$$Slope = \frac{\mu r \rho_c}{2 A \Delta P} \tag{12.11}$$

$$r = Slope \frac{2 \Delta P}{\mu \rho_c} \tag{12.12}$$

$$r = 5.818 \; s/cm^2 \frac{2 \times 500 mm \; Hg \times \frac{1.33 \times 10^3 g}{cm \; s^2 \; mm \; Hg}}{0.03 \; g/cm \; s \times 14 \; g/690 \; ml}$$

$$= 2.185 \times 10^9 \; cm/g \tag{12.13}$$

$$R_m = \frac{\Delta P}{\mu} \tag{12.14}$$

$$R_m = 4.7646 \; s/cm \frac{500 mm \; Hg \times \frac{1.33 \times 10^3 g}{cm \; s^2 \; mm \; Hg}}{0.03 \; g/cm \; s}$$

$$= 1.0565 \times 10^8 \; cm^{-1} \tag{12.15}$$

This is a typical value of R_m for a large-pore sized filter.

In order to determine the time required to obtain 5000 liters of filtrate, we are assuming the same specific cake resistance at 450 mm Hg pressure drop. Using the same line equation given below and substituting $x = V/A = 5000 \times 10^3 m^3 / 12 \times 10^4 \; cm^2 = 41.67$ results in,

$$y = 5.818x + 4.7646 \tag{12.16}$$

$$y = t/(V/A) = 5.818 \times 41.67 + 4.7646$$

$$\frac{t}{V/A} = 241.5063 \; s/cm \tag{12.17}$$

$$t = 2.7952 \; hrs \tag{12.18}$$

12.2 Crossflow Filtration

Crossflow filtration can be divided into two categories, namely, ultrafiltration if dissolved species such as proteins are filtered or microfiltration if suspended particles are present.

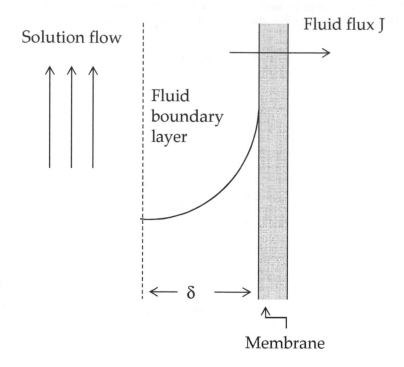

FIGURE 12.3
Schematic of concentration polarization in ultrafiltration

Ultrafiltration

In ultrafiltration where dissolved species are separated, a solution is flowing parallel to the membrane surface under pressure as shown in the schematic in Figure 12.3. For the solute which is rejected by the membrane, there is a concentration gradient created across the stagnant boundary layer next to the surface of the membrane. The ratio of concentration of solute at the membrane surface (c_m) to that of concentration in the bulk solution (c_b) is called concentration modulus indicative of concentration polarization.

At steady state, the rate of convective mass transfer of solute is equal to the rate of mass transfer diffusing away from the surface [119]

$$Jc = -D\frac{dc}{dx} \tag{12.19}$$

where J is the transmembrane fluid flux, c is the concentration of solute, and D is the diffusion coefficient of the solute. For a boundary layer of thickness δ as shown in Figure 12.3, the solution of the Equation 12.19 results in,

$$\frac{c_m}{c_b} \quad = \quad e^{\frac{J\delta}{D}} \tag{12.20}$$

$$= \quad e^{\frac{J}{k}} \tag{12.21}$$

where k is the mass transfer coefficient for which correlations have been developed for laminar and turbulent flow. For details, please see [6].

From Equation 12.21, it can be seen that the polarization modulus c_m/c_b is extremely sensitive to changes in J and k because of exponential functionality involved. For high molecular weight solutes (small k and D) and membrane with high solvent permeability (high J), concentration polarization becomes severe, resulting in precipitation of solute and formation of solids or gel layer on the membrane surface.

Microfiltration

Microfiltration is used when suspended particles are present. For suspended particles, the theory presented for dissolved solids earlier is only valid for very small suspended particles up to approximately 1 μm in size [120]. Two different theories are presented for larger size particles, namely, a shear induced diffusion theory [121] and an inertial lift theory [122]. Details of these theories are beyond the scope of this book. Please refer to [121] [123] for more information.

12.3   Types of Filtration Equipment

There is a large variety of conventional filtration equipment. One of the most common types of batch filter is the plate-and-frame filter (Figure 12.4). This filter consists of a number of filter chambers with medium-covered filter plates alternated with frames that provide space for the cake. The chambers are closed and tightened by a hydraulic ram or screw which pushes the plate and frame together. Other types of batch filters are horizontal or vertical leaf filter, tray filter, tube or candle filter, and Nutsche filter. These are all enclosed pressure filters.

Crossflow filtration membranes are available in a variety of configurations. These are hollow fiber, tubular, flat plate, spiral wound, and rotating. Some key characteristics of each module are given in Table 12.2 reproduced here from [6].

12.4   Summary

Filtration is a separation process which uses pressure differentials through a porous medium to separate suspended or dissolved solids. The two broad categories of filtration are conventional filtration where the fluid flow is perpendicular to the filter medium, and crossflow filtration where it is parallel to the medium. The filter media in conventional flitration are filter cloths, meshes or screens, while in crossflow

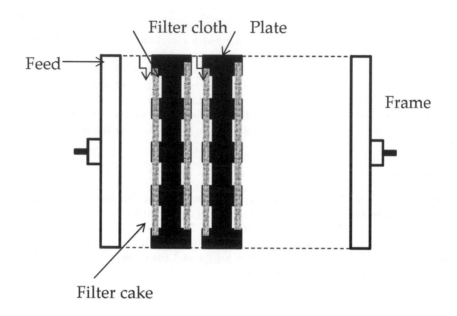

FIGURE 12.4
Schematic of plate-and-frame filter operation

TABLE 12.2
Comparison of key characteristics of crossflow membrane modules[6]

Module type	Channel spacing(cm)	Packing density (m^{-1})	Energy costs	Particulate plugging	Ease of cleaning
Hollow fiber	0.02-0.25	1200	Low	High	Fair
Tubular	1.0-2.5	60	High	Low	Excellent
Flat plate	0.03-0.25	300	Moderate	Moderate	Good
Spiral wound	0.03-0.1	600	Low	Very high	Poor-fair
Rotating	0.05-0.1	10	Very high	Moderate	Fair

filtration membranes are used as a filter medium. There are two kinds of crossflow filtration, namely, ultrafiltration for dissolved solids separation, and microfiltration for suspended solids separation. Conventional filtration operation obeys some form of Darcy's law. In crossflow filtration at equilibrium, the rate of convective mass transfer of solute towards the membrane surface is equal to the rate of mass transfer of solute by diffusion away from the membrane surface. Although, ultrafiltration follows this assumption, in microfiltration it is valid only for very small suspended particles. There are other theories developed for microfiltration for larger suspended particles. The most commonly used batch filter is the conventional plate-and-frame filter.

Notations

A	cross-sectional area [length^2]
c	concentration of solute [mass/length^3]
c_b	concentration of solute in the bulk [mass/length^3]
c_m	concentration of solute at the membrane surface [mass/length^3]
D	diffusion coefficient
e	voidage
J	transmembrane fluid flux
k	mass transfer coefficient
l	cake thickness [length]
r	specific resistance
R	resistance of porous bed
r'	empirical constant
s	empirical constant, cake compressibility factor
R_c	resistance of cake
R_m	resistance of medium
t	integration variable, time [time]
V	volume of filtrate [length^3]
x	integration variable, distance [length]

Greek Letters:

δ	boundary layer thickness [length]
μ	viscosity of filtrate [mass/length/time]
ρ_c	density of cake [mass/volume]

13

Batch Centrifugation

CONTENTS

In centrifugation,centrifugal force is used in the place of gravitational force for separation. The centrifugal force provides far greater rates of separation. In the centrifuge, the fluid is introduced on some form of rotating vessel and is rapidly accelerated. The relative motion between fluids separates different materials. Centrifugation is used for separating particles on the basis of their size or density, separating immiscible liquids of different densities, filtration of a suspension, breaking down of emulsions and colloidal suspensions, or for separation of gases, e.g., isotope separation in nuclear industry.

Centrifuges may be grouped into two distinct categories: those that utilize the principle of filtration and those that utilize the principle of sedimentation, both enhanced by the use of a centrifugal field. These two categories are analyzed separately below.

Batch centrifuges with imperforate bowls (sedimentation type) are used either for producing an accelerated separation of solid particles in a liquid, or for separating mixtures of two liquids. In the former case, the solids are deposited on the wall of the bowl and the liquid is removed through an overflow or skimming tube. The suspension is continuously fed in until a suitable depth of solids has been built up on the wall; this deposit is then removed either by hand or by a mechanical scraper. With the bowl mounted about a horizontal axis, solids are more readily discharged because they can be allowed to fall directly into a chute [3]. Perforated bowls are used when relatively large particles are to be separated from a liquid, as for example in the separation of crystals from a mother liquor. The mother liquor passes through the bed of particles and then through the perforations in the bowl. When the centrifuge is used for filtration, a coarse gauze is laid over the inner surface of the bowl and the filter cloth rests on the gauze. Space is thus provided behind the cloth for the filtrate to flow to the perforations[3].

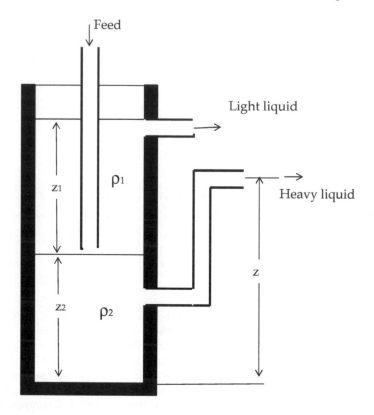

FIGURE 13.1
Schematic of the continuous gravity separator separating two immiscible fluids

13.1 Sedimentation in a Centrifugal Field

As stated earlier, batch centrifuges involve a simple bowl with centrifugal force. To understand the theory behind centrifuges, it is important to study the gravity settlers first. Figure 13.1 shows a schematic of the continuous gravity separator separating two immiscible fluids. At equilibrium, the hydrostatic pressure exerted by a height z of the heavier liquid is equal to a height z_2 of the heavier liquid and a height z_1 of the lighter liquid resulting in

$$z\rho_2 g \quad = \quad z_2\rho_2 g + z1\rho_1 g \tag{13.1}$$

$$z \quad = \quad z_2 + z1\frac{\rho_1}{\rho_2} \tag{13.2}$$

where ρ denotes density. Fluid 1 is lighter fluid and fluid 2 is heavier fluid.

A similar equation can be derived for batch centrifuge. A schematic of a batch

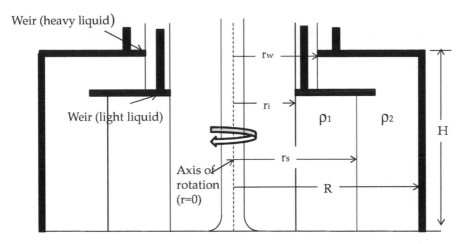

FIGURE 13.2
Schematic of the batch centrifuge separating two immiscible fluids

centrifuge bowl separating two immiscible liquids is shown in Figure 13.2. To derive this, we need to consider the pressure balance.

A force balance equation for a fluid in the rotating bowl rotating at angular velocity, ω leads to pressure gradient equation given below

$$\frac{\partial P}{\partial r} = \rho \omega^2 r \tag{13.3}$$

Integrating this equation provides an equation for pressure P exerted by the liquid on the walls of the bowl of radius R when the radius of the inner surface of the liquid is r_0 as

$$P = \frac{1}{2}\rho\omega^2(R^2 - r_0^2) \tag{13.4}$$

For the batch centrifuge bowl shown in Figure 13.2, the radius r_i of the weir of the lighter liquid correspond approximately to the radius of the inner surface of the liquid in the bowl. Then, the pressure developed by heavier liquid at the radius R with r_w, the radius of outer weir corresponding to the inner surface of the heavier liquid in the bowl is equal alone as it flows over the weir is equal to that due to the two liquids in the bowl. This results in the following equation

$$\frac{1}{2}\rho_2\omega^2(R^2 - r_w^2) = \frac{1}{2}\rho_2\omega^2(R^2 - r_s^2) + \frac{1}{2}\rho_1\omega^2(r_s^2 - r_i^2) \tag{13.5}$$

where r_s is the radius of the interface between the two liquids.

The retention time, t_r for the mixture with feed rate equal to Q is given by

$$t_r = \frac{V'}{Q} = \frac{V'}{Q_1 + Q_2} = \tag{13.6}$$

$$= \frac{\pi(R^2 - r_i^2)H}{Q} \tag{13.7}$$

where Q_1 is the feed flow of lighter liquid and Q_2 is the flow of heavier liquid. Since the residence time required by the two liquids is same. We can write

$$t_r = \frac{V_1}{Q_1} = \frac{V_2}{Q_2} \tag{13.8}$$

$$\frac{\pi(r_s^2 - r_i^2)H}{Q_1} = \frac{\pi(R^2 - r_s^2)H}{Q_2} \tag{13.9}$$

$$\frac{Q_1}{Q_2} = \frac{(r_s^2 - r_i^2}{(R^2 - r_s^2)} \tag{13.10}$$

From Equation 13.5 one can write

$$\frac{\rho_1}{\rho_2} = \frac{(r_s^2 - r_w^2)}{(r_s^2 - r_i^2)} \tag{13.11}$$

In practice the relative values of r_i and r_w are adjusted to modify the relative residence times of the individual phases to give extra separating time for the more difficult phase.

For separating suspended solids, the minimum retention time, t_r with a velocity u_0 and capacity factor Σ required for all particles of diameter greater than d to be deposited on the walls of the bowl through a liquid layer of thickness h (for $h \ll R$) is given by

$$t_r = \frac{18\mu h}{d^2(\rho_s - \rho)R\omega^2} \tag{13.12}$$

$$u_0 = \frac{d^2(\rho_s - \rho)R\omega^2 g}{18\mu} \tag{13.13}$$

$$\Sigma = \frac{R\omega^2 V'}{hg} \tag{13.14}$$

For h is comparable to R, the retention time, t_r and capacity factor, Σ are given by

$$t_r = \frac{18\mu h}{d^2(\rho_s - \rho)\omega^2} \ln\frac{R}{r_0} \tag{13.15}$$

$$\Sigma = \frac{\omega^2 V'}{hg \ln\frac{R}{r_0}} \tag{13.16}$$

13.2   Filtration in Centrifugal Field

In filtration centrifuge, a perforated bowl is used to permit removal of the filtrate. Figure 13.3 shows a schematic of filtration centrifuge operation where a bowl of radius R is used. The filter cake of thickness l is shown in the figure which increases and hence the resistance of the filter cake will increase as more solids are deposited. The fluid is introduced at such a rate that the inner radius r_0 remains constant. The radius of the interface between the cake and the suspension is r'.

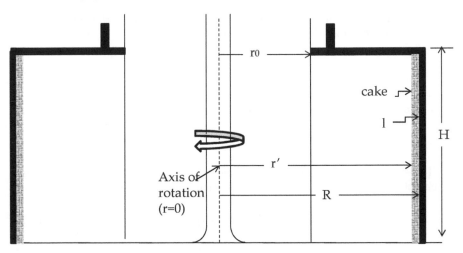

FIGURE 13.3
Schematic of the filtration centrifuge

Using Darcy's law one can write an equation for pressure drop $(\triangle P')$ across the filter cake of small thickness dl (please see Equations 12.1 and 12.4) as a function of velocity across the cake.

$$u_c = \frac{1}{A}\frac{dV}{dt} = \frac{\triangle P'}{\mu R_c} \tag{13.17}$$

$$= \frac{1}{\mu r \rho_c}\frac{-dP'}{dl} \tag{13.18}$$

In general the centrifugal force is large compared with the gravitational force, therefore, the filtrate will flow in an approximately radial direction, and will be evenly distributed over the axial length of the bowl. Therefore, Equation 13.18 can be written as

$$\frac{1}{2\pi r'H}\frac{dV}{dt} = \frac{1}{\mu r \rho_c}\frac{-dP'}{dl} \tag{13.19}$$

$$-dP' = \frac{\mu r \rho_c dl}{2\pi r'H}\frac{dV}{dt} \tag{13.20}$$

From the Figure 13.3, we know that $l = R - r'$ and, therefore, $dl = -dr'$. Substituting the value of dl in Equation 13.20 and integrating for pressure drop across the cake results in

$$-\triangle P' = \frac{\mu r \rho_c}{2\pi H}\frac{dV}{dt}\int_{r'=r'}^{r'=R}\frac{dr'}{r'} \tag{13.21}$$

$$-\triangle P' = \frac{\mu r \rho_c}{2\pi H}\frac{dV}{dt}\ln\frac{R}{r'} \tag{13.22}$$

If we want to include the resistance of the medium also in the equation, then the pressure drop across the medium can be written as

$$-\Delta P'' = \frac{\mu r \rho_c}{2\pi H} \frac{dV}{dt} \frac{L}{R} \tag{13.23}$$

where L is equivalent resistance of the medium in terms of length L.

The total pressure drop then is given by

$$-\Delta P = -\Delta P' - \Delta P'' = \frac{\mu r \rho_c}{2\pi H} \frac{dV}{dt} (\ln \frac{R}{r'} + \frac{L}{R}) \tag{13.24}$$

If v is the bulk volume of incompressible cake deposited by the passage of unit volume of filtrate, then we can write the relation between r' and V as follows.

$$v dV = 2\pi r' H dr' \tag{13.25}$$

$$\frac{dV}{dt} = \frac{2\pi r' H}{v} \frac{dr'}{dt} \tag{13.26}$$

Substituting in Equation 13.24 and integrating from $r' = r'$ to $r' = R$ as t goes from 0 to t results in

$$\frac{v(-\Delta P)t}{\mu r \rho_c} = (\frac{1}{4} + \frac{L}{2R})(R^2 - r'^2) + \frac{1}{2} r' \ln \frac{r'}{R} \tag{13.27}$$

From Equation 13.4, we know that $-\Delta P = \frac{1}{2}\rho \omega^2 (R^2 - r_0^2)$. Therefore,

$$t = \frac{\mu r \rho_c}{2 v \rho \omega^2 (R^2 - r_0^2)} (\frac{1}{4} + \frac{L}{2R})(R^2 - r'^2) + \frac{1}{2} r' \ln \frac{r'}{R} \tag{13.28}$$

13.3   Summary

Centrifugation uses centrifugal force to separate particles on the basis of their size or density, to separate immiscible liquids of different densities, filter a suspension or break down an emulsion and colloidal suspension, or to separate isotopes. Centrifuges can be classified into two types: ones which utilize filtration and ones that use sedimentation for separation. Theoretical analysis changes depending on the type of classification.

Notations

A	area [length^2]
d	size of the particle [length]
g	gravitational acceleration [length/time^2]

h	liquid layer thickness [length]
H	height [length]
Q	feed rate of mixture [mass/time]
Q_1	feed rate of liquid 1 [mass/time]
Q_2	feed rate of liquid 2 [mass/time]
r_i	radius of inner weir [length]
r_0	radius of inner surface of liquid [length]
r_s	radius of the interface between the two liquids [length]
r_w	radius of outer weir [length]
R	radius of the bowl [length]
t	integration variable, time [time]
t_r	retention time [time]
u_0	velocity [length/time]
V'	volume of the liquid mixture [volume]
V_1	volume of the liquid 1 [length^3]
V_2	volume of the liquid 2 [length^3]
z	height of liquid [length]
z_1	height of liquid 1 [length]
z_2	height of liquid 2 [length]

Greek Letters:

Σ	capacity factor
ρ	density of liquid[mass/vol]
ρ_s	density of suspended solids [mass/vol]
ω	angular velocity [angle/time]

14

Batch Scheduling and Planning

CONTENTS

Saadet Ulas Acikgoz"" UOP, A Honeywell Company,"" 25 E. Algonquin Rd.,""
Des Plaines, IL 60017

Batch plants produce multiple products in a series of processing steps called a batch recipe. These products are processed in a certain period of time as dictated by market demands and requirements. Contrary to continuous chemical processes, for batch processing, the market demand is not defined for a certain quantity of a single product, but instead a range of products for which the demand is continuously changing and new products are introduced on a regular basis. For production of a single product in a continuous chemical processing plant, the sequence of process tasks are decided at the design stage for a specific market capacity and the equipment is designed specifically to perform those process tasks. In a batch production environment, often the same resources or equipment are used to process multiple products and the allocation of process tasks to equipment is constantly changing to meet the varying market demands and requirements. In order to reduce inventory, lower manufacturing costs, meet market demands in a timely manner, and maximize the efficiency of available resources, batch scheduling and planning is used to assign process equipment to produce multiple products in a certain amount of time following a sequence of process tasks for each product.

Consider the example of a batch processing plant which produces 3 products as shown in the flow diagram below in Figure 14.1. According to this flow diagram, Products A and C are processed in 3 steps: mixing, reaction, and separation. On the other hand Product B is processed in 5 steps: heating, reaction 1, reaction 2, separation, and drying. Products A and C can be processed in either Reactor 1 or Reactor 2. Separator and reactors are shared equipment used for the processing of all products.

The processing times for each step are shown in Table 14.1 for Products A, B, and C. Let us determine the minimum processing time for all products. If we can produce all 3 products simultaneously, in the time it takes to produce Product B which has a longer processing time, our minimum processing time will be equivalent to the total processing time for Product B which is 13 hours. Figure 14.2 shows the production schedule of Products A, B, and C according to this schedule, Products A and B

179

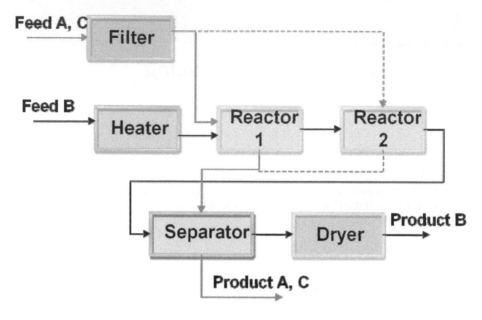

FIGURE 14.1
Simple batch process flow diagram for Products A, B, and C

are produced simultaneously in Reactors 1 and 2. Then Reactor 2 is used for the processing of Product B and Reactor 1 is used for the processing of Product C. This schedule makes use of the fact that both reactors can be used for the processing of all products and allocating them to minimize the processing time for all. The graphic representation of the production schedule as shown in Figure 14.2 is called a Gantt chart. Gantt charts were developed by Henry Gantt and are commonly used as tools in project management and scheduling of project activities.

The example shown above is a multipurpose, multiproduct batch plant consisting of multiple stages. These types of plants are called job-shop plants in literature. In job-shop scheduling problems, different products may follow different paths for production. Another example of a multiple product batch plant is a plant where all

TABLE 14.1
Processing times for Products A, B, and C in each equipment

Processing times (hr)	Product A	Product B	Product C
Filter	1		0.5
Heater		1	
Reactor 1	6	6	4
Reactor 2		3	
Separator	1	2	1
Dryer		1	
Total time	8	13	5.5

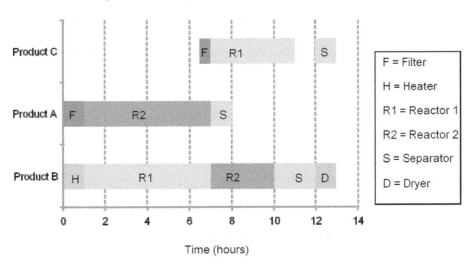

FIGURE 14.2
Gantt chart for the production schedule of Products A, B, and C

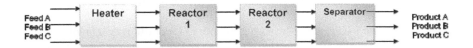

FIGURE 14.3
An example of a flowshop plant

products follow the same production sequence. This is called a flowshop plant. An example of this is shown in Figure 14.3.

14.1   Types of Scheduling Problems

Batch scheduling problems have been classified according to process topology, equipment assignment and connectivity, inventory storage policies, material transfer, batch sizes and processing times, demand patterns, changeovers, resource and time constraints, and costs[124]. In a batch plant, the general short term scheduling problem can be defined by these parameters [125] :

1.   A set of N products or product batches to be produced
2.   A set of M processing units
3.   A sequence for each product in which operations are to be performed
4.   A set of processing times for each product on processing units. The batch

processing times may be fixed and unit dependent but may also be variable based on the batch size.

5. A matrix of transfer times for each product from each equipment item

6. A matrix of setup times or costs between every pair of products in each equipment item. These setup times can be unit and product dependent and may be different for each sequence.

7. Constraints on the production order for some products. There may be constraints due to demand patterns, where production targets have to be met, or resource constraints such as labor or utilities.

8. The nature of intermediate storage between processing stages. Four different storage options are defined:

   - Unlimited intermediate storage (UIS)

   - Finite intermediate storage (FIS)

   - No intermediate storage (NIS)

   - Zero wait or no wait (ZW or NW)

9. The structure of processing network: single stage or multi-stage, single unit or multiple parallel units in each stage. For multi-stage units, the structure can be classified as multipurpose (job-shop) or a multiproduct (flowshop) plant. Examples of these are shown in Figure 14.1 and Figure 14.3.

10. A suitable performance or cost criterion to be optimized. The objective functions that have been used as performance criteria are listed below:

    - Makespan: Total time required to produce all products utilizing the available equipment and resources of the plant

    - Flow time: The time required to pass completely through the process averaged over all products

    - Maximum flow time among products

    - Mean tardiness of the schedule, where tardiness is defined as the difference between the delivery date of the product and its due date.

    - Maximum tardiness

    - Changeover or setup cost

14.2   Role of Optimization in Batch Scheduling

The objective of the short term scheduling problem is to decide on which sequence or order the N products must be produced and what the starting and finishing times for each product should be on the M processing units. This is a highly complex combinatorial optimization problem for many industrial plants. An example of an industrial scheduling problem will be presented in the last section of this chapter. All industrial scheduling problems are known to belong to a set of decision problems called NP-complete based on computational complexity theory. The most important characteristic of these problems is that there is no known quick solution to them. The time required to solve the problem using any currently known algorithms increases rapidly as the size of the problem grows. Several strategies can be applied for the solution of batch scheduling problems [126].

- Strategy I: Simplifying/approximating the problem to be solved to make it easier. The batch scheduling problem is modified by changing the constraints, parameters, and objective function until it can be solved by a scheduling algorithm.

- Strategy II: Using an exact algorithm. The scheduling problems which can be formulated as mixed integer linear programming (MILP) problems can be solved using available optimization packages or a branch and bound (BAB) strategy. A MILP problem is an optimization problem with linear objective function and constraints but some of the variables are integer or binary. The scheduling problem can be formulated using recurrence relations, which are a set of expressions to define the start and finish times for each product in each processing stage and storage unit. Floudas and Lin [127] presented a review of MILP based approaches in solving process scheduling problems. MILP based approaches may require an unreasonable amount of time based on the problem size so they are often paired with the other strategies for solving these types of problems.

- Strategy III: Using proababilistic methods or heuristics. Heuristic algorithms generate suboptimal solutions that are based on heuristic reasoning and are used when the problem size is large and a MILP or branch and bound algorithm is not able to solve the problem. For example simulated annealing is a heuristic combinatorial optimization method based on the ideas from statistical mechanics. It uses the analogy to physical annealing, where all atomic particles arrange themselves in a lattice formation to minimize their energy. The objective function becomes the energy of the system which is minimized by generating random perturbations to the system. The moves are accepted according to the Metropolis criteria. Evolutionary algorithms such as genetic algorithms have also been applied to solve job-shop scheduling problems. Genetic algorithms are based on the theory of evolution. The evolution usually starts from a population of randomly generated individuals and happens in generations. In each generation, the fitness of every individual in the population is evaluated, multiple individuals are stochastically selected from the current population (based on their fitness), and modified (recombined and possibly randomly mutated) to form a new population. The new population is then used in the next iteration of the algorithm. With each generation, the optimization problem is evolved to a better solution. The algorithm terminates when either a maximum number of generations has been reached, or a satisfactory fitness level has been achieved for the population. If the algorithm has terminated due to a maximum number of generations, a satisfactory solution may or may not have been reached. A discussion on simulated annealing method and genetic algorithms are presented in Diwekar [70].

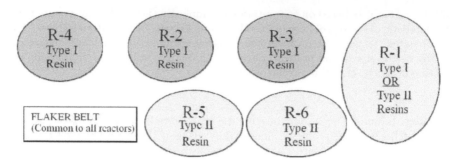

FIGURE 14.4
Resin plant equipments

14.3 Industrial Scheduling Problem

An industrial scheduling problem is presented here to illustrate the concepts discussed in this chapter. The industrial problem presented here is from a resin manufacturing plant and it involves optimizing the production schedule in order to enhance the manufacturing process performance. All production in this plant is in batch mode and three kinds of resins are produced.

There are six reactors and a flaker belt located in the resin plant. The equipment used in the plant are given in Figure 14.4. Reactors 1 through 4 are horizontal reactors mainly used to produce Type I resins. Reactor 1 is the only reactor that is used for both Type I and Type II resins. Reactors 5 and 6 are vertical reactors used for Type II resins only. After going through the reactors, the final product is either flaked or a resin solution is made. Flaking is a process where the molten resin is discharged onto the flaker belt and cooled by spraying water. As a result the molten resin solidifies and it is bagged as 50-lb or bulk bags. Some of these resins are not flaked and instead resin solutions are made by mixing resin and ink oils.

The main bottleneck of this plant is the flaker belt. Since there is only one flaker belt, the batch which has come to specifications might be kept inside the reactor depending on the batches competing for the flaker belt. Therefore the major cost associated with flaking is this tied up reactor time. Some of the formulas are less stable than others, therefore hold times for these formulas are minimized and the first priority is given to them. Making solutions takes as much time as flaking but since the resin solutions bypass the flaker belt, they don't have any resource conflicts.

In this problem the top 10 products which account for 80% of production in the resin plant are considered.

A literature survey was performed in order to find a formulation suitable for this scheduling problem. The problem described above is similar to a flexible flowshop scheduling problem with limited intermediate storage. This is a machine environment which consists of a number of stages in series with a number of machines in parallel at each stage. A job has to be processed at each stage on only one of these machines. This kind of a problem is also referred to as a multiprocessor flowshop or hybrid flowshop scheduling problem [128]. When there is finite intermediate storage (FIS) or no intermediate storage (NIS) between the machines, this means that if a given machine finishes the processing of a given job, the job cannot proceed to the next machine if it is busy and it must be kept on the machine.

It is possible to find many papers related to flexible flowshop scheduling. Recently, Wang [129] has reviewed different solution strategies for these kinds of problems. There are three kinds of solutions: optimum, heuristics, and artificial intelligence. The optimum approach uses a branch and bound method which guarantees optimal solutions. When it is computationally expensive to use an exact solution technique, researchers have often used heuristics to solve this problem. The final solution technique is artificial intelligence. The AI techniques commonly employed to solve flexible flowshop scheduling problems are neural networks, genetic algorithms, tabu search, and simulated annealing.

Our problem consists of two stages in series: the reactor and flaker belt. In the first stage, there are 5 parallel reactors, whereas in the second stage there is only one flaker belt. Since there is zero intermediate storage between these stages, the resin which is processed must remain in the reactor if the flaker belt is busy.

The only difference of our problem from a standard flexible flowshop scheduling problem is the fact that our reactors in the first stage are not identical. Not all of the products can be processed in all of the reactors. For example, hydrocarbon resins are only produced in Reactors 5 and 6, which are vertical reactors, whereas rosin resins are only produced in Reactors 1 through 4, which are horizontal reactors. This makes our problem more complex than a standard flexible flowshop scheduling problem because we also have to consider machine eligibility.

Two papers were found addressing similar problems. Ruiz and Maroto [130] have developed a genetic algorithm for hybrid flow shops with sequence dependent setup times and machine eligibility. In this paper, a similar problem is considered in the ceramic tile industry where all of the tile types follow the same flow in the shop but not all products can be processed by all machines at a given production stage and parallel machines at each stage are unrelated. The second paper is authored by Sawik [131], where mixed integer linear programming is used for scheduling flexible flow lines with limited intermediate buffers. The basic mixed integer programming formulations are enhanced in this paper by considering alternative processing routes for each product. Both formulations were suitable for our problem, however the mixed integer programming formulation was chosen in this case and this problem was solved using commercially available software AIMMS with XA solver.

The problem formulation considers a flexible flow line with alternative processing routes for each product. For each route and product an ordered set of stages is given that the product must visit. The product must be processed by at most one machine in each of such stages and it bypasses the other stages that are not in the route. This formulation also considers limited or no intermediate storage space between the stages.

The mixed integer linear programming formulation was developed by Sawik [131] and is presented below:

Model FFR: Scheduling flexible flow line with limited intermediate buffers and alternative routings

$$\text{Maximize} \quad C_{max} \tag{14.1}$$
$$x$$

subject to

Assignment constraints for stages with parallel processors:

$$\sum_{j \in J} x_{i,j,k,r} = z_r,$$
$$\dots k \in K, r \in R_k, i \in I_r :\mid J_i \mid> 1, \tag{14.2}$$

$$\sum_{k \in K} \sum_{r \in R_k} p_{i,k,r} x_{i,j,k,r} \leq \sum_{k \in K} \sum_{r \in R_k} \frac{p_{i,k,r} z_r}{\mid J_i \mid} + \min p_{i,k,r},$$
$$k \in K, r \in R_k, i \in I, j \in J_i :\mid J_i \mid> 1, \tag{14.3}$$

Product completion constraints:

$$c_{f,k,r} \leq p_{f,k,r} z_r, \dots k \in K, r \in R_k \tag{14.4}$$
$$c_{i,k,r} - c_{i-1,k,r} \leq p_{i,k,r} z_r,$$
$$k \in K \quad, r \in R_k, i \in I_r : I_r > f_r \tag{14.5}$$

Product noninterference constraints for stages with single processors:

$$c_{i,k,r} + Q(2 + y_{k,l} - z_r - z_s) \leq p_{i,k,r} z_r \tag{14.6}$$
$$k, l \in K, r \in R_k, s \in R_l, i \in I_k \bigcap I_l \mid J_i \mid= 1$$
$$c_{i,l,s} + Q(3 - y_{k,l} - z_r - z_s) \leq d_{i,k,r} + p_{i,l,s} z_s \tag{14.7}$$
$$k, l \in, r \in R_k, \quad s \in R_l \quad, i \in I_k \bigcap I_l : k < l \ \& \mid J_i \mid= 1$$

Product noninterference constraints for stages with parallel processors:

$$c_{i,k,r} + Q(2 + y_{k,l} - x_{i,j,k,r} - x_{i,j,l,s}) \geq d_{i,l,s} + p_{i,k,r} z_r \tag{14.8}$$
$$k, l \in, r \in R_k, s \in R_l, \quad i \in I_k \quad \bigcap I_l \ \& \mid J_i \mid= 1$$

$$c_{i,l,s} + Q(3 - y_{k,l} - x_{i,j,k,r} - x_{i,j,l,s}) \leq d_{i,k,r} + p_{i,l,s} z_s \tag{14.9}$$
$$k, l \in K, r \in R_k, s \in R_l, \quad i \in I_k \quad \bigcap I_l : k < l \ and \mid J_i \mid= 1$$

No store constraints:

$$c_{i,k,r} = d_{i-1,k,r} + p_{i,k,r} z_r, \dots k \in K, r \in R_k, i \in I_r : I_r > f_r \tag{14.10}$$

Completion time constraints:

$$c_{i,k,r} \leq Q z_r, \dots k \in K, r \in R_k, i \in I_r \tag{14.11}$$
$$d_{i,k,r} \leq Q z_r, \dots k \in K, r \in R_k, i \in I_r \tag{14.12}$$
$$c_{l,k,r} = d_{l,k,r}, \dots k \in K, r \in R_k, \tag{14.13}$$
$$c_{l,k,r} \leq C_{max}, \dots k \in K, r \in R_k, \tag{14.14}$$
$$C_{max} \geq \sum_{k \in K} \sum_{r \in R_k} \frac{p_{i,k,r} z_r}{\mid J_i \mid} + \min \sum_{h \in l; h \neq l} p_{i,k,r}, \dots i \in I_r \tag{14.15}$$

Route selection constraints:

$$\sum_{r \in R_k} z_r \;=\; 1, \ldots k \in K \tag{14.16}$$

Variable elimination constraints:

$$x_{i,j,k,r} \;=\; 0, \ldots k \in K, r \in R_k, i \in I_r, j \in J_i : |\, J_i \,| = 1 \tag{14.17}$$
$$y_{k,l} \;=\; 0, \ldots k, l \in K : k \geq l \tag{14.18}$$

Variable non-negativity and integrality consraints:

$$c_{i,k,r} \;\geq\; 0, \ldots k \in K, r \in R_k, i \in I_r, \tag{14.19}$$
$$d_{i,k,r} \;\geq\; 0, \ldots k \in K, r \in R_k, i \in I_r, \tag{14.20}$$
$$x_{i,j,k,r} \;\in\; \{0,1\}, \ldots k \in K, r \in R_k, i \in I_r, j \in J_i \tag{14.21}$$
$$y_{k,l} \;\in\; \{0,1\}, \ldots k, l \in K \tag{14.22}$$

In this formulation Equation 14.1 represents the objective function which is the schedule length to be minimized. Constraint (Equation 14.2) ensures that in every stage that has parallel processors, each product is assigned to exactly one processor, if the stage is in the selected processing route. Otherwise, the product is not assigned to this stage. Constraint (Equation 14.3) is for equalizing the work load assigned to each parallel processor. Constraint (Equation 14.4) ensures that each product is processed in the first stage in its route selected and constraint (Equation 14.5) ensures that each product is also processed in all the downstream stages in their routes. Constraints (Equation 14.6) and (Equation 14.7) are product noninterference constraints for stages with single processors. For parallel processors, constraints (Equation 14.8) and (Equation 14.9) are used. For a given sequence of products and processing routes selected, only one constraint is active, and only if both product k and l are assigned to the same processor. Equation 14.10 states that processing of each product in every stage starts immediately after its departure from the previous stage of the selected route. Constraints (Equation 14.11) and (Equation 14.12) impose upper bounds on completion and departure times. Equation 14.13 ensures that each product at the last stage of their processing route leave the system. Constraint (Equation 14.14) defines the maximum completion time and Equation 14.15 imposes a lower bound on it. Finally Equation 14.16 ensures that only one processing route is selected for each product.

Results of a simple scheduling problem for a flexible flow line with no buffers

This example problem was solved in order to test the program written in AIMMS using XA solver. The flexible flow line with no buffers is made up of $m = 3$ processing stages. The number of processors in each stage is $n1 = 1$, $n2 = 2$ and $n3 = 1$. There are 5 products and the processing routes for three-stage flow line with no buffer are given below:

$$k = 1: \quad R1 = [1,2], \quad I1 = [1,3], \quad I2 = [2,3]$$
$$k = 2: \quad R2 = [3], \quad I2 = [1,3],$$
$$k = 3: \quad R3 = [4], \quad I3 = [2,3],$$

TABLE 14.2

Processing times for each product

i (stage)	k (product)	r (route)	$p_{i,k,r}$ processing time
1	1	1	4
2	1	2	4
3	1	1	2
3	1	2	2
1	2	3	2
3	2	3	5
2	3	4	4
3	3	4	7
1	4	5	5
3	4	5	7
2	5	6	4
3	5	6	7

$$k = 4: \quad R4 = [5], \quad I4 = [1,3],$$
$$k = 5: \quad R5 = [6], \quad I5 = [2,3],$$

The processing times for each product $p_{i,k,r}$ are given in Table 14.2.

This problem was solved in AIMMS with the XA solver. The minimum schedule length was found as 30. The completion and departure times for each product are given in Tables 14.3 and 14.4. The Gantt chart for this example is given in Figure 14.5.

In this section, the resins plant data was used for the MILP problem. The current schedule of the plant was taken which consists of a sequence of 17 batches. In these 17 batches, the plant's top 10 products are processed. These products are named P1, P2, P3, P4, P5, P6, P7, P8, P9, and P10. For confidentiality reasons, the commercial names of these products will not be disclosed here. The average processing times (cycle times) for these products are given in Table 14.5. The last three products were not considered in the optimization problem because these products are not flaked and instead made solutions. Therefore, the flaker belt is not allocated for these products. For each product there are 2 possible processing routes. Products P1 and P2 can go through Reactors 1 to 4 and the flaker belt. Products P3 and P4 can only be processed in Reactors 1, 5 and 6 and the flaker belt. Since Reactor 1, and the flaker belt are common to all of these products, they were considered as stages that the products must be processed according to the route chosen. Therefore, the available routes are:

$$Route-1: \quad Reactor1, \quad Flaker Belt(stage1, stage4) \quad I1 = [1,4]$$
$$Route-2: \quad Reactors2-4, \quad Flaker Belt(stage2, stage4) \quad I2 = [2,4]$$
$$Route-3: \quad Reactors5-6, \quad Flaker Belt(stage3, stage4) \quad I3 = [3,4]$$

For products P1, P2, P8, and P9 routes 1 and 2 are allowed. For products P3, P5, and P10 routes 1 and 3 are allowed. (R1=[1,2], R2=[1,2], R8=[1,2], R9=[1,2],

TABLE 14.3

Completion times for each product

i (stages)	k (products)	r (routes)	$c_{i,k,r}$ completion times
1	p1	route-1	11
1	p2	route-1	2
1	p4	route-1	7
2	p3	route-2	21
2	p5	route-2	14
3	p1	route-1	30
3	p2	route-1	7
3	p3	route-2	28
3	p4	route-1	14
3	p5	route-2	21

TABLE 14.4

Departure times for each product

i (stages)	k (products)	r (routes)	$d_{i,k,r}$ departure time
1	p1	route-1	28
1	p2	route-1	2
1	p4	route-1	7
2	p3	route-2	21
2	p5	route-2	14
3	p1	route-1	30
3	p2	route-1	7
3	p3	route-2	28
3	p4	route-1	14
3	p5	route-2	21

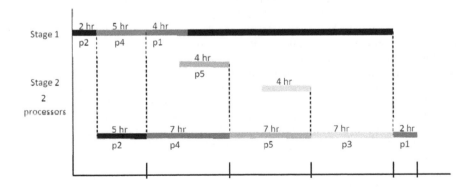

FIGURE 14.5

Gantt chart for the example problem

TABLE 14.5

Processing times for each product

i (stage)	k (product)	r (route)	$p_{i,k,r}$ processing time
1,2	P1	1,2	40
4	P1	1,2	8
1,2	P2	1,2	39
4	P2	1,2	8
1,3	P3	1,3	32
4	P3	1,3	8
1,3	P5	1,3	32
4	P5	1,3	8
1,3	P8	1,2	36
4	P8	1,2	8
1,3	P9	1,2	36
4	P9	1,2	8
1,3	P10	1,3	31
4	P10	1,3	8
3	P4	solution	21
1,3	P6	solution	28
1,3	P7	solution	33

R3=[1,3], R5=[1,3], R10=[1,3]) In stage 2, there are 3 parallel processors (reactors), where the products can be processed. In stage 3, there are 2 parallel processors (reactors), $(n1 = 1, n2 = 3, n3 = 2, n4 = 1)$. The first processing stage in each route is given as $f1 = 1, f2 = 2$, and $f3 = 3$. The last processing stage for each route is stage 4, $l1 = 4, l2 = 4, l3 = 4, l4 = 4$. In this production schedule, 3 batches of P1, 3 batches of P2, 1 batch of P3, 2 batches of P5, 1 batch of P8, 2 batches of P9, and 2 batches of P10 will be produced. The total number of batches is 17 $(v = 17)$ and the number of processing stages is 4 $(m = 4)$. The number of routes is 3 $(w = 3)$. The processing times $p_{i,k,r}$ for each product at each stage and route is given in Table 14.5. The final piece of information to solve this problem is Q. The large number Q which should not be less than schedule length was chosen as 500. This number should be chosen carefully because it affects the program.

This problem was solved in AIMMS in order to find the optimal schedule length using an MILP solver. The optimal solution was found as 144 hours. The problem characteristics of this MILP program are presented in Table 14.6. This solution was reached in 30 seconds. It took 40000 iterations and 3847 nodes to obtain this solution. After this solution was found, the program continued till 1000 seconds time limit and this solution was proven optimal within this time limit. The departure time for each product from each stage is given in Table 14.7. Also, the binary variable $x_{i,j,k,r}$ which shows which product is assigned to which processor is given in Table 14.8. The suffix at the end of each product name specifies the batch number. For example P11 means, product 1, batch 1.

The current schedule of the plant is given in a Gantt chart in Figure 14.6. The Gantt chart for this schedule according to Tables 14.7 and 14.8 is given in Figure 14.7. For the current schedule, the schedule length is 181 hours. In the Gantt chart

TABLE 14.6
Problem characteristics of the MILP program

Problem	Time	Total number of nodes	Total number of iterations	# Constraints	#Variables	#Non-zeros	Optimal
Resin Plant MILP	1000 sec	121,218	970,516	1714	416	718	144

TABLE 14.7

Departure times for each product

i (stage)	k (products)	r (route)	$d_{i,k,r}$ departure time
st1	P31	route-1	71
st2	P13	route-2	120
st2	P11	route-2	87
st2	P12	route-2	95
st2	P23	route-2	128
st2	P21	route-2	55
st2	P22	route-2	79
st2	P81	route-2	47
st2	P92	route-2	136
st2	P91	route-2	39
st3	P101	route-3	31
st3	P102	route-3	104
st3	P51	route-3	63
st3	P52	route-3	112
st4	P101	route-3	39
st4	P102	route-3	112
st4	P13	route-2	128
st4	P11	route-2	95
st4	P12	route-2	103
st4	P23	route-2	136
st4	P21	route-2	63
st4	P22	route-2	87
st4	P31	route-1	79
st4	P51	route-3	71
st4	P52	route-3	120
st4	P81	route-2	55
st4	P92	route-2	144
st4	P91	route-2	47

TABLE 14.8
Assignment of processors to products

i (stage)	j (processor)	k (product)	r (route)	$x_{i,j,k,r}$ binary variable
st2	Reactor2	P13	route-2	1
st2	Reactor2	P22	route-2	1
st2	Reactor2	P91	route-2	1
st2	Reactor3	P12	route-2	1
st2	Reactor3	P21	route-2	1
st2	Reactor3	P92	route-2	1
st2	Reactor4	P11	route-2	1
st2	Reactor4	P23	route-2	1
st2	Reactor4	P81	route-2	1
st3	Reactor5	P51	route-3	1
st3	Reactor5	P102	route-3	1
st3	Reactor6	P52	route-3	1

for the current schedule, the products which are made solutions are also shown. These are products P1, P6, and P7. All three of these products are processed in Reactor 6 in this schedule. These products are also shown on the Gantt chart for the optimal schedule, although they were not considered in the optimization problem. These three products are shown in gray color and they were inserted in the optimal schedule after the MILP problem was solved. Since these products are not processed in the flaker belt, they do not change the outcome of the optimization problem. The optimal length of the schedule is the same. This procedure reduces the number of variables and constraints for this MILP problem.

It can be seen that using this optimization method, the plant will benefit from a shorter schedule length. The optimal schedule length for this 17 batch sequence was found as 144, whereas according to the current schedule, these products are processed in 181 hours according to average cycle times. This schedule results in 20% time savings. A separate analysis can be performed in order to predict the cost savings for the plant from this shorter schedule, taking into account the demand and sales price for each product.

The solution found by MILP program is deterministic and it is subject to change according to changing cycle times for these products. The plant should be scheduled considering the uncertainties in processing times (cycle times) which vary according to various factors such as raw materials, the quality control time, which is the time required for the product to come to specifications inside the reactor and material charging accuracy. Since the plant operates in batch mode, it is easier to deal with process variations. Therefore it is essential to formulate this problem accounting for these uncertainties. Consideration of these uncertainties will make this problem more complex. The problem characteristics were previously shown in Table 14.6. Uncertainties will add more nodes to this problem and increase the computational intensity. The artificial intelligence methods mentioned in the introduction of this report, such as genetic algorithms, neural networks, or simulated annealing can be useful methods to solve this problem in the face of uncertainties. This is also going

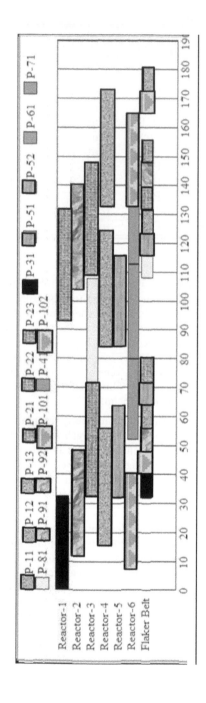

FIGURE 14.6

Gantt chart for the resin plant scheduling problem, current schedule of the plant (average cycle times)

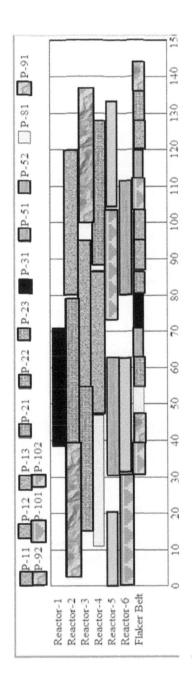

FIGURE 14.7
Gantt chart for the resin plant scheduling problem, optimal solution found by the MILP program

to help in identifying configurational changes for the plant which might decrease the sensitivity of the product schedule to variations in cycle times for the products.

14.4   Summary

Scheduling and planning are important for lowering manufacturing costs, reducing inventory, meeting market demands in a timely manner, and maximizing the efficiency of available resources in a multi-product multi-purpose batch manufacturing environment. A review of the types of scheduling problems and optimization methods in batch scheduling has been presented. Batch scheduling problems can be classified as multi-product (flowshop) or multi-purpose (jobshop) scheduling and a suitable performance criterion such as makespan (total time required to produce all products) or flow time is optimized. Scheduling problems can be solved using exact algorithms (MILP or branch and bound), proababilistic and/or heuristic algorithms (simulated annealing or genetic algorithms) or a combination of both when solution of the exact algorithms take an unreasonable amount of time due to problem size. An industrial scheduling problem was presented here to illustrate the concepts presented in this chapter.

Notations

Indices:

i	processing stage, $i \in I = \{1, \ldots, m\}$
j	processor in stage $i, j \in J_i = \{1, \ldots, n_i\}$
k	product, $i \in I = (1, \ldots, \gamma)$
r	processing route, $r \in R = (1, \ldots, \omega)$

Input Parameters:

R_k	set of processing routes available for product k,
I_r	ordered set of processing stages for route r,
	$I_r = \{f_r, \ldots, i_r, \ldots, l_r\} \subset I$
f_r	first processing stage in route r
l_r	last processing stage in route r
n_i	number of parallel processors in stage i
$p_{i,k,r}$	processing time for product k in stage i and route r
m	number of processing stages
$gamma$	number of products
ω	number of processing routes
Q	large number not less than schedule length

Decision Variables:

C_{max}	schedule length
$c_{i,k,r}$	completion time of product k in stage i and route r [time]
$d_{i,k,r}$	departure time from stage i for product k and route r[time]
$x_{i,j,k,r}$	binary variable =1, if product k is processed via route r is assigned to processor in j otherwise 0
$y_{k,l}$	binary variable =1, if product k precedes product l; otherwise 0
z_r	binary variable=1, if processing route r is selected

15

Batch Process Simulation

CONTENTS

Demetri Petrides*, Douglas Carmichael, and Charles Siletti
Intelligen, Inc. Scotch Plains, NJ, USA
Corresponding author: dpetrides@intelligen.com

15.1 Introduction

Batch process simulation is a computer modeling technique used for the design, analysis, and optimization of batch manufacturing processes. Batch process manufacturing is practiced in industries that produce low-volume, high-value products such as pharmaceuticals, fine chemicals, biochemicals, food, consumer products, etc. Most batch manufacturing facilities are multiproduct plants that produce a variety of products.

Previous chapters described modeling of individual operations, such as distillation, crystallization, filtration, etc. Batch process simulation combines multiple processing steps in order to model a complete batch process or even a multiproduct plant.

This chapter is divided into three sections. Section 15.1 provides introductory information and explains the benefits of batch process simulation. Section 15.2 focuses on the detailed modeling of single batch processes. This type of analysis is typically performed by chemical engineers engaged in the development and optimization of such processes. The concepts are illustrated with an example that deals with the

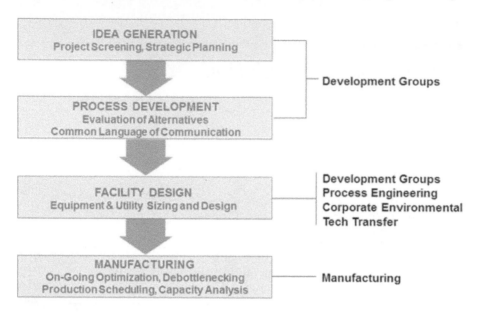

FIGURE 15.1

Uses of batch process simulation throughout the lifecycle of a product

manufacturing of an active pharmaceutical ingredient (API). Section 15.3 focuses on modeling of multiproduct plants. Information is provided on approaches and tools, and the key concepts are illustrated with an example that deals with manufacturing of polymer resins. The focus of Section 15.3 is on production planning and scheduling issues since that is the primary objective of multiproduct plant modeling. At the conclusion of this chapter, readers should have a good understanding of the steps involved with creating and analyzing a batch process model, the typical uses of single-product and multiproduct simulation models, the main outputs from these models, and the key benefits of batch process simulation.

15.1.1   Benefits of Batch Process Simulation

Figure 15.1 displays the benefits of process simulation and scheduling tools throughout the life-cycle of product development and commercialization [132, 133, 134]. At the early stages of idea generation, process simulation is primarily used for screening and evaluating potential projects in order to determine which ones to move forward with. In process development, simulation tools are used to evaluate alternative processing scenarios from an economics, cycle time reduction, and environmental point of view. Cost-of-goods analysis facilitates identification of the critical steps and aspects of a process, and this information is used to guide subsequent research and development work. Capital cost estimation facilitates decisions related to in-house manufacturing versus outsourcing.

When a process is ready to move from development to manufacturing, process simulation facilitates technology transfer and process fitting. A detailed computer model provides a thorough description of a process in a way that can be readily understood and adjusted by the recipients. Process adjustments are commonly required when a new process is moved into an existing facility whose equipment is not ideally sized for the new process. The simulation model is then used to adjust batch sizes, define cycling of certain steps (for equipment that cannot handle a batch in one cycle), etc. Furthermore, if a new plant is being designed or an existing plant is being retrofitted, the model can be used to determine the size of the main equipment as well as the capacity of the supporting utility systems that supply steam, electricity, purified water, etc. Estimation of the labor requirement is another capability of such tools.

Multiproduct plant modeling and scheduling tools also play an important role in manufacturing. They facilitate capacity analysis and long term planning, and enable production scheduling on a day-to-day basis in a way that does not violate constraints related to the limited availability of equipment, labor, utilities, inventories of materials, etc. Production scheduling tools close the gap between ERP/MRP II tools and the plant floor [135, 136]. Production schedules generated by ERP (Enterprise Resource Planning) and MRP II (Manufacturing Resource Planning) tools are typically based on coarse process representations and approximate plant capacities. As a result, solutions generated by these tools may not be feasible, especially for multiproduct facilities that operate at high capacity utilization. This often leads to late orders that require expediting and/or to large inventories required to maintain customer responsiveness. "Lean manufacturing" principles, such as just-in-time production, low work-in-progress (WIP), and low product inventories cannot be implemented without good production scheduling tools that can accurately estimate capacity.

Sensitivity analyses are greatly facilitated by process simulation tools as well. The objective of such studies is to evaluate the impact of critical parameters on various key performance indicators (KPIs), such as production cost, cycle times, and plant throughput. If there is uncertainty for certain input parameters, sensitivity analysis can be supplemented with Monte Carlo simulation to quantify the impact of uncertainty.

15.2 Detailed Modeling of Single Batch Processes

Modeling and analysis of integrated batch processes is facilitated by process simulators. Simulation tools have been used in the chemical and petrochemical industries since the early 1960s. Simulators for these industries have been designed to model continuous processes and their transient behavior for process control purposes. Continuous modeling applications, however, cannot effectively model the sequential operations associated with batch and semi-continuous processes. Such processes are best modeled with batch process simulators that account for time-dependency and sequencing of events. BATCHES from Batch Process Technologies (a Purdue University spin-off located in West Lafayette, IN) was the first simulator specific to batch processing. It was commercialized in the mid-1980s. All of its operation models are dynamic, and simulation always involves integration of differential equations over a period of time. In the mid-1990s, Aspen Technology (Burlington, MA) introduced

Batch Plus (now called Aspen Batch Process Developer), a recipe-driven simulator that targeted batch pharmaceutical processes. Around the same time, Intelligen (Scotch Plains, NJ) introduced SuperPro Designer. The initial focus of SuperPro Designer was on bioprocessing. Over the years, its scope has been expanded to support modeling of fine chemicals, pharmaceuticals, food processing, consumer products, and other types of batch/semi-continuous processes.

The steps involved in the development of a batch process model will be illustrated with a simple process that represents the manufacturing of an active pharmaceutical ingredient (API) for skin care applications. SuperPro Designer will be used for the development and analysis of this model.

15.2.1   Model Creation Steps

The first step in building a simulation model is always the collection of information about the process. Engineers rely on draft versions of process descriptions, flow diagrams, and batch sheets from past runs, which contain information on material inputs, operating conditions, etc. Reasonable assumptions and approximations are then made for missing data. The following information is required for modeling a batch process:

- Processing steps, their durations, performance parameters, and sequencing

- Equipment available to processing steps

- Materials consumed by or generated by the process

- Other resource requirements (e.g., labor, utilities, etc.)

It is a good practice to build the model step-by-step, checking and verifying the results of the evolving model before additional complexity is added [137, 138]. The registration of materials (pure components and mixtures) is usually the first step in building the model. Next, the flow diagram (see Figure 15.2) is developed by adding the required "unit procedures" (process steps) and joining them with material flow streams. Operations are then added to unit procedures and their operating conditions and performance parameters are specified.

In SuperPro Designer, the representation of a batch process model is loosely based on the ISA S-88 standards for batch recipe representation [139]. A batch process model is in essence a batch recipe that describes how to make a certain quantity of a specific product. The set of operations that comprise a processing step is called a "unit procedure" (as opposed to a unit operation which is a term used for continuous processes). The individual tasks contained in a procedure are called "operations." A unit procedure is represented on the screen with a single equipment-like icon. Figure 15.3 displays the dialog through which operations are added to a vessel unit procedure. On the left-hand side of that dialog, the program displays the operations that are available in the context of a vessel procedure; on the right-hand side, it displays the registered operations (Charge Quinaldine, Charge Chlorine, Charge Na_2CO_3, Agitate, etc.). The two-level representation of operations in the context of unit procedures enables users to describe and model batch processes in detail.

For every operation within a unit procedure, the simulator includes a mathematical model that performs material and energy balance calculations. Based on the results of the material and energy balances, it then performs equipment-sizing

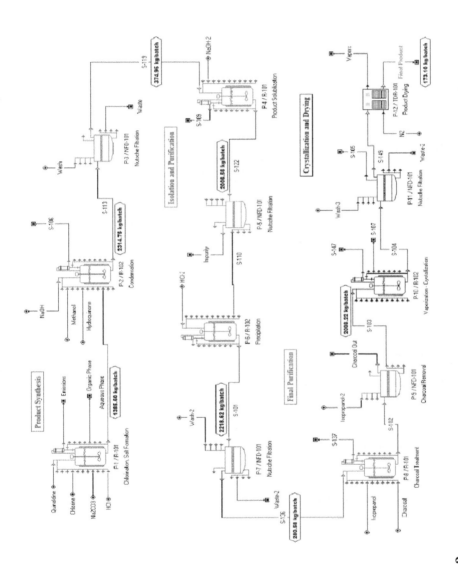

FIGURE 15.2
Flow diagram of an API process

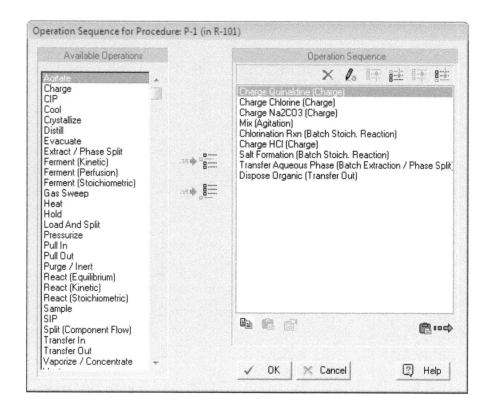

FIGURE 15.3
The operations associated with the first unit procedure ("P-1") of Figure 15.2

calculations. If multiple operations within a unit procedure dictate different sizes for a certain piece of equipment, the software reconciles the different demands and selects an equipment size that is appropriate for all operations. The equipment is sized so that it is large enough, and hence not overfilled during any operation, but it is no larger than necessary (in order to minimize capital costs). If the equipment size is known and specified by the user, which is the case when processes that utilize existing facilities are modeled, the simulator checks to make sure that the vessel is not overfilled. In addition, the tool checks to ensure that the vessel contents do not fall below a user-specified minimum volume (e.g., a minimum stirring volume) for applicable operations.

In addition to material balances, equipment sizing, and cycle time analysis, the simulator can be used to carry out cost-of-goods analysis and project economic evaluation. The sections that follow present illustrative outputs of the above.

Having developed a good model using a process simulator, the user may begin experimenting on the simulator with alternative process setups and operating conditions. This has the potential of reducing the costly and time-consuming laboratory and pilot plant efforts. Of course, the GIGO (garbage-in, garbage-out) principle applies to all computer models. If critical assumptions and input data are incorrect, the outcome of the simulation will be incorrect as well. A certain level of model verification is therefore necessary after the development of a batch process model. In its simplest form, a review of the results by an experienced engineer can play the role of verification.

15.2.2   API Manufacturing Example

Process Description and Design Basis

Figure 15.2 displays the flow diagram of the simple API manufacturing process analyzed in this example. It corresponds to case "C" of the SynPharm example of SuperPro Designer which is available online [140]. A production suite is dedicated to this process which initially includes two 3,800 L reactors (R-101 and R-102), one 2.5 m^2 Nutsche filter (NFD-101), and a 10 m^2 tray dryer (TDR-101). The objective is to produce 27,000 kg of active ingredient per year at a cost of no more than \$350/kg.

The process is divided into four sections: 1) Product Synthesis, 2) Isolation and Purification, 3) Final Purification, and 4) Crystallization and Drying. A flowsheet section in SuperPro Designer is simply a group of unit procedures (processing steps). If you open the file (SPhr9_0c.spf) in SuperPro Designer, you will see that the unit procedures of each section are differentiated by color (green, blue, purple, and black for sections one, two, three, and four, respectively).

The process includes a total of 12 unit procedures. The first reaction step (procedure P-1) involves the chlorination of quinaldine. Quinaldine is dissolved in carbon tetrachloride (CCl_4) and reacts with gaseous Cl_2 to form chloroquinaldine. The conversion of the reaction is around 98% (based on amount of quanaldine fed). The generated HCl is then neutralized using Na_2CO_3. The stoichiometry of these reactions follows:

$$Quinaldine + Cl_2 \quad \rightarrow \quad Chloroquinaldine + HCl$$
$$Na_2CO_3 + HCl \quad \rightarrow \quad NaHCO_3 + NaCl$$
$$NaHCO_3 + HCl \quad \rightarrow \quad NaCl + H_2O + CO_2$$

The small amounts of unreacted Cl_2, generated CO_2, and volatilized CCl_4 are vented. The above three reactions occur sequentially in the first reactor vessel (R-101). Next, HCl is added in order to produce chloroquinaldine-HCl. The HCl first neutralizes the remaining $NaHCO_3$ and then reacts with chloroquinaldine to form its salt, according to the following stoichiometries:

$$NaHCO_3 + HCl \quad \rightarrow \quad NaCl + H_2O + CO_2$$
$$Chloroquinaldine + HCl \quad \rightarrow \quad Chloroquinaldine.HCl$$

The small amounts of generated CO_2 and volatilized CCl_4 are vented. The presence of water (added with HCl as hydrochloric acid solution) and CCl_4 leads to the formation of two liquid phases. Then the small amounts of unreacted quinaldine and chloroquinaldine are removed with the organic phase. The chloroquinaldine-HCl remains in the aqueous phase. This sequence of operations (including all charges and transfers) requires 14.5 h.

After removal of the unreacted quinaldine, the condensation of chloroquinaldine and hydroquinone takes place in reactor R-102 (procedure P-2). First, the salt chloroquinaldine-HCl is converted back to chloroquinaldine using $NaOH$. Then, hydroquinone reacts with $NaOH$ and yields hydroquinone-Na. Finally, chloroquinaldine and hydroquinone-Na react and yield the desired intermediate product. Along with product formation, roughly 2% of chloroquinaldine dimerizes and forms an undesirable by-product impurity. This series of reactions and transfers takes 13.3 h. The stoichiometry of these reactions follows:

$$Chloroquinaldine.HCl + NaOH \quad \rightarrow \quad NaCl + H_2O + Chloroquinaldine$$
$$2Chloroquinaldine + 2NaOH \quad \rightarrow \quad 2H_2O + 2NaCl + Impurity$$
$$Hydroquinone + NaOH \quad \rightarrow \quad H_2O + Hydroquinone.Na$$
$$Chloroquinaldine + Hydroquinone.Na \quad \rightarrow \quad Product + NaCl$$

Both the Product and Impurity molecules formed during the condensation reaction precipitate out of solution and are recovered using a Nutsche filter (procedure P-3, filter NFD-101). The product recovery yield is 90%. The sum of the filtration, wash, and cake removal times in P-3 is 6.4 h.

Next, the Product/Impurity cake recovered by filtration is added into a $NaOH$ solution in reactor R-101 (procedure P-4). The Product molecules react with $NaOH$ to form Product-Na, which is soluble in water. The Impurity molecules remain in the solid phase, and are subsequently removed during procedure P-5 in filter NFD-101. The Product remains dissolved in the liquors. Procedure P-4 takes about 10 h, and procedure P-5 takes approximately 4 h.

Notice that the single filter (NFD-101) is used by several different procedures. The two reactors are also used for multiple procedures during each batch. Please note that the equipment icons in Figure 15.2 represent unit procedures (processing steps), as opposed to unique pieces of equipment. The procedure names (P-1, P-3, etc.) below the icons refer to the unit procedures, whereas the equipment tag names (R-101, R-102, etc.) refer to the actual physical pieces of equipment. The process flow diagram in SuperPro Designer is essentially a graphical representation of the batch recipe that displays the execution sequence of the various steps.

After the filtration in procedure P-5, the excess $NaOH$ is neutralized using HCl and the Product-Na salt is converted back to Product in Reactor R-102 (procedure P-6). Since the Product is insoluble in water, it precipitates out of solution. The Product is then recovered using another filtration step in procedure P-7. The Product recovery yield is 90%. The precipitation procedure takes roughly 10.7 h, and the filtration takes about 5.7 h. The recovered Product cake is then dissolved in isopropanol and treated with charcoal to remove coloration. This takes place within reactor R-101 in procedure P-8. After charcoal treatment, the solid carbon particles are removed using another filtration step in (procedure P-9). The times required for charcoal treatment and filtration are 15.9 h and 5 h, respectively.

In the next step (procedure P-10), the solvent is distilled off until the solution is half its original volume. The temperature of the solution is lowered and the product crystallizes in the same vessel with a yield of 97%. The crystalline product is recovered with a 90% yield using a final filtration step (procedure P-11). The distillation and crystallization steps take a total of 18.3 h, and filtration requires roughly 3.3 h. The recovered product crystals are then dried in a tray dryer (procedure P-12, TDR-101). This takes an additional 15.6 h. The amount of purified Product generated per batch is 173.1 kg.

SuperPro Designer and other batch process simulators provide multiple mathematical models for each type of operation. For instance, options for chemical reactions include stoichiometric, kinetic, equilibrium, etc. The selection of the appropriate option depends on data available and the type of questions that the model is supposed to answer. The operation durations in such tools are either specified by the user or calculated by the tool based on performance parameters (e.g., the pumping rate of a transfer operation or the vaporization rate of a distill operation). The relative timing (sequencing) of operations is specified by the user. Figure 15.4 displays the Gantt chart of this process, which enables users to visualize the execution of a batch as a function of time. The bar at the top (displayed by a crossed hatch pattern) represents the duration of the entire process (91.64 h). The dark bars underneath represent procedures and the white bars represent operations of the associated procedures. Each bar in Figure 15.4 has a label to the right of it to identify the specific activity that it represents, and its duration is displayed in parenthesis. The grid on the left-hand-side of the chart displays the same information in tabular form. The [+] and [-] buttons in front of the procedure names in the Task column enable users to display or hide the operations of a procedure. Only Procedures P-6 and P-7 are expanded (displaying their operations) in Figure 15.4. Notice the parallel execution of the "TRANSFER-OUT-1" in P-6 and the "Prod Isolation" (filtration operation) in P-7. These two operations represent the simultaneous material transfer from P-6 (R-102) to P-7 (NFD-101) and filtration in P-7 (NFD-101).

Table 15.1 displays the raw material requirements in kg per batch and per kg of main product (MP = purified Product). Note that around 54.3 kg of raw materials (solvents, reagents, etc.) are used per kg of product produced. Thus the product to raw material ratio is only 1.84%, an indication that large amounts of waste are generated by this process.

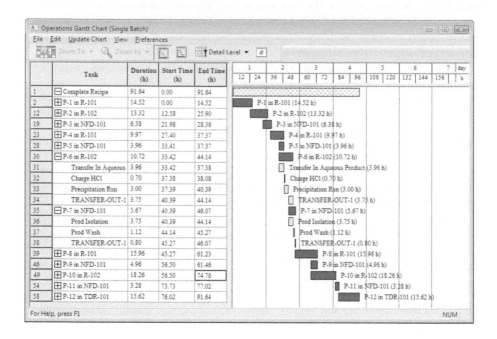

FIGURE 15.4

Gantt chart of the API manufacturing process

TABLE 15.1

Raw material requirements

Material	kg/batch	kg/kg MP
Carb. TetraCh	497.31	2.87
Quinaldine	148.63	0.86
Water	3,621.44	20.92
Chlorine	89.52	0.52
Na_2CO_3	105.06	0.61
HCl (20% w/w)	357.44	2.07
$NaOH$ (50% w/w)	204.52	1.18
Methanol	553.26	3.20
Hydroquinone	171.45	0.99
Sodium Hydroxide	74.16	0.43
HCl (37% w/w)	217.57	1.26
Isopropanol	2,232.14	12.90
Charcoal	15.85	0.09
Nitrogen	1,111.49	6.42
TOTAL 9,399.84 54.30		

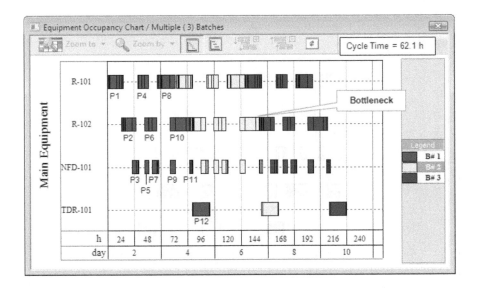

FIGURE 15.5
Equipment occupancy chart for three consecutive batches

Cycle Time Reduction and Throughput Increase

Figure 15.5 displays the Equipment Occupancy chart for three consecutive batches of the process (each color represents a different batch, and the individual procedures associated with Batch 1 are labeled). The process batch time is approximately 92 h. This is the total time between the start of the first operation of a batch and the end of the last operation of that batch. However, since most of the equipment items are utilized for shorter periods within a batch, a new batch can be initiated every 62.1 h. This is known as the "minimum cycle time" of the process. Multiple bars of the same color on a given line (e.g., for R-101, R-102, and NFD-101) represent reuse (sharing) of equipment by multiple procedures. White space between bars represents idle time. The equipment with the least idle time between consecutive batches is the time (or scheduling) bottleneck (R-102 in this case), which determines the maximum number of batches per year. Its occupancy time (62.1 h) is the minimum possible time between consecutive batches. In contrast, the occupancy time of R-101 is 61.3 h, so there is a small gap between batches in this unit.

Scheduling and cycle time analysis in the context of a simulator is fully process-driven and the impact of process changes can be analyzed in a matter of seconds. For instance, the impact of an increase in batch size (which affects the duration of charge, transfer, filtration, distillation, and other scale-dependent operations) on the recipe cycle time and the maximum number of batches can be seen immediately. Due to the many interacting factors involved with even a relatively simple process, simulation tools that allow users to describe their processes in detail, and to quickly perform what-if analyses, can be extremely useful.

If this production line operated around the clock for 330 days a year (7,920 h) with its minimum cycle time of 62.1 h, its maximum annual number of batches

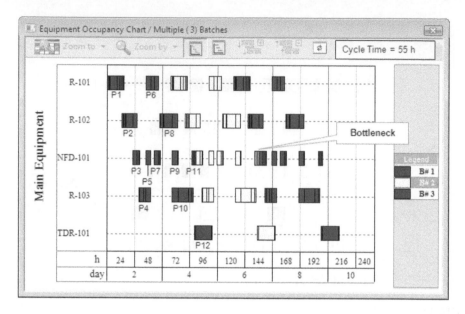

FIGURE 15.6
Equipment occupancy chart for the case with three reactors

would be 126, leading to an annual production of 21,810 kg of API (126 batches x 173.1 kg/batch), which is less than the project's objective of 27,000 kg. Please note that the maximum annual number of batches is less than 7920/62.1 because of the limited overlap of activities during the start and end of the campaign.

Since the process already operates at its maximum possible batch size (based on the equipment capacities), the only way to increase production is by reducing the process cycle time and thus increasing the number of batches per year. The cycle time can be reduced through process changes or by addition of extra equipment. However, major process changes in regulated industries such as pharmaceuticals usually require regulatory approval and are avoided in practice. As a result, addition of extra equipment may be required in order to achieve cycle time reduction. Since R-102 is the current bottleneck, addition of an extra reactor can shift the bottleneck to another unit (in this case, the new bottleneck is the filter NFD-101). Figure 15.6 displays the effect of the addition of an extra reactor (R-103). As before, the specific procedures associated with Batch 1 are labeled on the chart. Please note that under the new conditions each reactor handles only two procedures instead of three. As a result, the batch cycle time of each reactor is reduced.

The addition of R-103 reduces the cycle time of the process to 55 h, resulting in 143 batches per year and annual throughput of 24,753 kg. Under these conditions the bottleneck shifts to NFD-101. Since the annual throughput is still below the desired amount of 27,000 kg/year, addition of an extra Nutsche filter to eliminate the current bottleneck is the next logical step. Figure 15.7 shows the results of that scenario. Under these conditions, the first Nutsche filter (NFD-101) is used for the first three filtration procedures (P-3, P-5, and P-7) and the second filter (NFD-102)

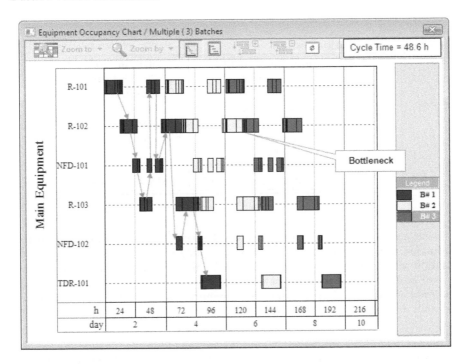

FIGURE 15.7
Equipment occupancy chart for the case with three reactors and two filters

handles the last two filtration procedures (P-9 and P-11). This reduces the recipe cycle time to 48.6 h, allowing for up to 162 batches per year and maximum annual throughput of 28,042 kg, which meets the production objective of the project. The gray arrows of Figure 15.7 represent the flow of material through the equipment for the first batch.

Debottlenecking projects that involve installation of additional equipment provide an opportunity for batch size increases that can lead to substantial throughput increase. More specifically, if the size of the new reactor (R-103) is selected to accommodate the needs of the most demanding vessel procedure (based on volumetric utilization) in a way that shifts the batch size bottleneck to another procedure, then that creates an opportunity for batch size increase. Additional information on debottlenecking and throughput increase options can be found in the literature [140, 141].

Cost Analysis

Estimation of capital and operating costs are important for a number of reasons. If a company lacks a suitable manufacturing facility with available capacity to accommodate a new product, it must decide whether to build a new plant or outsource the production. Building a new plant is a major capital expenditure and a lengthy process. To make a decision, management must have information on the capital in-

TABLE 15.2

Key economic evaluation results

Total Capital Investment	$24.1 million
Plant Throughput	27,000 kg/year
Manufacturing Cost	$9.3 million/year
Unit Production Cost	$345/kg
Selling Price	$600/kg
Revenues	$16.2 million/year
Gross Margin	42.4%
IRR (after taxes)	17.7%
NPV (for 7% discount interest)	$18.4 million

vestment and the time required to complete the facility. If the plan is to outsource production, the company should still do a cost analysis and use it as a basis for negotiation with contract manufacturers. A sufficiently detailed computer model can be used as the basis for the discussion and negotiation of the terms. Contract manufacturers usually base their estimates on requirements of equipment utilization, materials and labor per batch, which is information that can be provided by the model.

SuperPro Designer estimates equipment cost using built-in cost correlations that are based on data derived from a number of vendors and literature sources. In addition, users have the flexibility to enter their own data and correlations for equipment cost estimation. The fixed capital investment is estimated based on equipment cost using various multipliers, some of which are equipment specific (e.g., installation cost) while others are plant specific (e.g., cost of piping, buildings, etc.). This approach is described in detail in the literature [6, 142]. The rest of this section provides a summary of the cost analysis results for this example process. Table 15.2 shows the key economic evaluation results for this project. Key assumptions for the economic evaluations include: 1) a new production suite will be built and dedicated to the manufacturing of this product; 2) the entire direct fixed capital is depreciated linearly over a period of twelve years; 3) the project lifetime is 15 years, and 4) 27,000 kg of final product is produced per year.

For a plant of this capacity, the total capital investment is around $24.1 million (Table 15.2). The unit production cost is $345/kg of product, which satisfies the project's objective for a unit cost of under $350/kg. Assuming a selling price of $600/kg, the project yields an after-tax internal rate of return (IRR) of 17.7% and a net present value (NPV) of $18.4 million (assuming a discount interest of 7%).

Figure 15.8 breaks down the manufacturing cost. The facility-dependent cost, which primarily accounts for the depreciation and maintenance of the plant, is the most important item accounting for 40.8% of the overall cost. This is common for high-value products that are produced in small facilities. This cost can be reduced by manufacturing the product at a facility whose equipment has already been depreciated. Raw materials are the second most important cost item accounting for 29.6% of the total manufacturing cost. Furthermore, if we look more closely at the raw material cost breakdown, it becomes evident that quinaldine, hydroquinone, and isopropanol make up more than 70% of this cost (see Table15.3). If a lower-priced

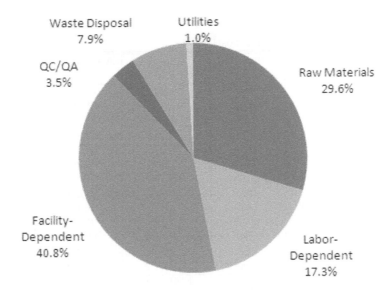

FIGURE 15.8

Manufacturing cost breakdown

quinaldine vendor could be found, the overall manufacturing cost would be reduced significantly.

Labor is the third most important cost item, accounting for 17.3% of the overall cost. The program estimates that 12 operators are required to run the plant around the clock, supported by 4 QC/QA scientists along with administrative and management personnel. The labor cost can be reduced by increasing automation or by locating the facility in a region of lower labor cost.

Accounting for Uncertainty and Variability

Process simulation tools used for batch process design, debottlenecking, and cost estimation typically employ deterministic models. In other words, they model the "average" or "expected" situation commonly referred to as the base case or most likely scenario. Sensitivity analysis can help determine the expected range of performance with respect to key process parameters. That can be supplemented with Monte Carlo simulation for uncertain input variables for which probability distributions are available. Monte Carlo simulation provides a practical means of quantifying the risk associated with uncertainty in process parameters [143].

In batch processing, uncertainty can emerge in operational or market-related parameters. Processing rates, duration of activities, material purchasing and product selling prices are common uncertain variables. Such uncertainty affects the throughput of a plant, the profitability and other key performance indicators (KPIs). In Monte Carlo simulation, numerous scenarios of a model are simulated by repeatedly picking values from a user-defined probability distribution for the uncertain variables. Those values are used in the model to calculate and analyze the outputs in a

TABLE 15.3

Raw material requirements and costs

Bulk Material	Unit Cost ($/kg)	Annual Amount (kg)	Annual Cost ($)	%
Carb.TetraCh	0.80	85,909	68,727	3.61
Quinaldine	32.00	25,676	821,629	43.14
Water	0.10	625,589	62,559	3.28
Chlorine	3.30	15,464	51,030	2.68
Na_2CO_3	6.50	18,148	117,963	6.19
$NaOH$ (50% w/w)	0.15	35,330	5,299	0.28
Methanol	0.24	95,573	22,938	1.2
Hydroquinone	4.00	29,617	118,468	6.22
Sodium Hydroxide	2.00	12,811	25,622	1.35
HCl (37% w/w)	0.17	37,585	6,389	0.34
Isopropanol	1.10	380,234	418,258	21.96
Charcoal	2.20	2,738	6,024	0.32
Nitrogen	1.00	179,787	179,787	9.44
TOTAL	1,544,461	1,904,694	100	

statistical way in order to quantify risk. The outcome of this analysis is the estimation of the confidence by which desired values of KPIs can be achieved. Inversely, the analysis can help identify the input parameters with the greatest effect on the bottom line and the input value ranges that minimize output uncertainty.

For a batch process modeled in SuperPro Designer, Monte Carlo simulation can be performed by combining Monte Carlo formulation tools, such as Oracle Crystal Ball from Oracle Corporation (Redwood Shores, CA) and @Risk from Palisade Corporation (Ithaca, NY) with SuperPro. Both Oracle Crystal Ball and @Risk are Microsoft Excel add-in applications that enable users to assign probability distributions to input parameters, manage the Monte Carlo simulation calculations, record the values of output variables, and analyze them statistically. The communication between Oracle Crystal Ball or @Risk and SuperPro is accomplished using Visual Basic for Applications (VBA) scripts written in Excel. A detailed Monte Carlo simulation example based on a variation of this process that utilizes Oracle Crystal Ball and SuperPro Designer is available in the literature [144].

15.3   Modeling of Multiproduct Batch Plants

The focus of batch process simulation, as described earlier in this chapter, is on the detailed modeling of a single batch process. The output of such models includes thorough material and energy balances, equipment sizing and selection, economic evaluation, and cycle time analysis. Such models facilitate scale up/down calculations, technology transfer, and process fitting. The primary objective of such models is to optimize processes under development. Modeling of multiproduct batch plants, in contrast, is focused on evaluating the interactions among multiple processes running concurrently in a plant. In most applications of multiproduct plant modeling, the processes are quite well defined. As a result, such models place less emphasis on process design calculations and more emphasis on timing and utilization of resources such as equipment, materials, utilities, labor, inventories of materials, etc. Typical applications of multiproduct batch plant modeling include:

- Estimation of Resource Requirements
- Capacity Analysis
- Production Planning and Scheduling

Estimation of Resource Requirements

During the design of new facilities or the debottlenecking and retrofit of existing ones, multiproduct plant models are used to estimate resource requirements in order to achieve a certain production volume within a given time period. Typical resources include main equipment (e.g., reactors, filters, dryers, etc.), utilities (e.g., steam, purified water, and electrical power), auxiliary equipment (e.g., shared pipe segments, transfer panels, cleaning equipment), various types of labor, etc. Engineering companies involved in the design of new facilities frequently use such models [134].

Capacity Analysis

This is the inverse of the resource requirement problem. The objective is to determine the production capability of an existing plant given a typical product mix and time horizon. Capacity analysis models must account for resource sharing among the processes of the various products, consider downtime for holidays, equipment preventive maintenance, and changeover activities, and consider other constraints that limit production capacity.

Production Planning and Scheduling

Production planning is related to capacity analysis. Its objective is to assign resources and estimate the start times for campaigns and batches and to determine the time horizon for producing specific quantities of a certain mix of products. The time horizon is usually weeks to months but it may extend to years for certain industries (e.g., pharmaceuticals). Typical outputs of such models include allocation of production lines and main equipment to specific product campaigns. In addition, planning models are used to estimate the demand for materials that need to be purchased, especially those requiring long lead times. Production scheduling focuses on the short term allocation of resources to various production activities. The time horizon is usually days to weeks but it may extend to months for certain industries (e.g., bio-pharmaceuticals).

The above applications require plant models at different levels of detail. Long term capacity analysis requires the least detail in recipe representation. Typically, the recipe of a batch process is represented with a small number of activities occupying key equipment that is likely to limit production. Long term planning utilizes models with a detail similar to those of capacity analysis. Short term planning and scheduling require more detailed models that account for the utilization of all main equipment and the critical auxiliary equipment (i.e., those with high utilization that are likely bottlenecks). Very detailed models are required for cycle time reduction and debottlenecking studies aimed at increasing the production of existing manufacturing lines and facilities. Such models should account for the occupancy of all equipment (main and auxiliary) as well as all other resources (e.g., utilities and materials) whose supply is limited by generation and inventory systems. In general, as the scope of the problem increases (in terms of number of products and time horizon) it is advisable to use simplified models for recipe representations so that the problem remains manageable.

15.3.1   Approaches to Modeling of Multiproduct Batch Plants

The approaches and tools utilized for modeling and scheduling of multiproduct batch plants vary widely. A possible grouping of the approaches based on tools utilized follows:

- Spreadsheet Tools
- Batch Process Simulation Tools
- Discrete Event Simulation Tools
- Mathematical Optimization Tools
- Recipe-Based Scheduling Tools

Spreadsheet Tools

People manually color Excel cells to represent the equivalent of equipment occupancy charts for consecutive batches. Some users have implemented scripts that color cells based on batch recipe descriptions as well. Very little sophistication is required for the user. However, this approach is time consuming, cannot be very detailed, and cannot be readily updated to account for delays and equipment failures. Nevertheless, it is probably the most common approach at the present.

Batch Process Simulation Tools

BATCHES and Aspen Batch Process Developer can be used to model multiple batches of multiple products. However, they take a long time to generate solutions because they do detailed material and energy balances for all the simulated batches. Furthermore, these tools cannot easily account for equipment failures, delays, work shift patterns, downtime for equipment maintenance, holidays, etc. Consequently, they cannot be used for day-to-day scheduling of multiproduct plants.

Discrete Event Simulation Tools

Discrete-event simulation (DES) is a popular technique for modeling of multiproduct batch plants. A series of dispatch rules govern which tasks may begin or end depending on the state and time. Dispatch rules and state calculations must often be custom-coded to capture many batch processes. An advantage of DES is the ability to perform stochastic modeling by accounting for the uncertainty and variability of certain input parameters. However, such tools cannot be used to schedule manufacturing facilities on a day-to-day basis because they cannot represent a specific plant situation and the user cannot take control. Established DES tools include Pro-Model from ProModel Corporation (Orem, UT), Arena and Witness from Rockwell Automation (Milwaukee, WI), and Extend from Imagine That (San Jose, CA).

Mathematical Optimization Tools

This approach is explained in detailed in Chapter 14. Optimization tools attempt to generate the best feasible solution by reshuffling campaigns and batches within the constraints set by the user. Such tools have found good applications in the industry for supply chain optimization and long-term planning that are based on simplified recipe (process) representations. Generating good solutions for problems that utilize detailed recipes with many constraints is quite challenging with such tools because they require very sophisticated users for the formulation of the problem. In many cases, even the solution algorithm must be tailored to the formulation of the problem. This is a highly specialized skill [145]. Established mathematical optimization tools with production planning and scheduling capabilities include SAP APO from SAP AG (Walldorf, Germany), IBM ILOG Plant PowerOps from IBM Corporation (Armonk, New York), Aspen Plant Scheduler from Aspen Technology (Burlington, MA), etc.

Recipe-Based Scheduling Tools

These tools functionally bridge the gap between batch process simulators and mathematical optimization scheduling tools. A batch process is represented as a recipe which describes a series of steps, the resources they require, and their relative timing and precedence. A production run is represented as a prioritized set of batches where each batch is one execution of a recipe with specific resources. Each batch is assigned to resources in priority order. Batches may be scheduled forward from a release date or backward from a due date. The scheduling algorithm generates feasible solutions that do not violate constraints related to the limited availability of resources. Partial optimization is attempted through the minimization of the production makespan. Such tools do not perform material and energy balance calculations around operations, but they keep track of the consumption and generation of materials, utilities, labor, and other resources. Also, they do not size equipment but they consider equipment capacity during resource allocation. For instance, if a vessel is too large or too small for a specific task, it will be ignored by the resource allocation algorithm. Such tools understand calendar time and consider work shift patterns. Equipment and facility downtime to account for preventive maintenance and holidays is readily specified. A number of recipe-based scheduling tools are available on the market, such as Preactor from Preactor International (Wiltshire, UK) and Orchestrate from Production Modeling (Coventry, UK). The majority target applications in the discrete manufacturing industries (assembly-type of production). SchedulePro from Intelligen (Scotch Plains, NJ) is probably the only tool of this type that focuses on batch chemical manufacturing. Discrete manufacturing tools tend to have a flat task-resource representation that is less convenient for process applications. For example a process that involves charging material to a vessel, blending, and transfer to another vessel requires at least three separate tasks (charging, blending, and transfer) along with special constraints to ensure that the vessel resource is the same for each task and that no other tasks may be assigned to the vessel from the start of the charge to the end of the transfer. Alternatively, all three tasks could

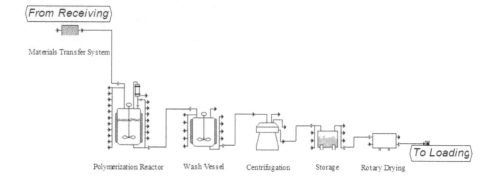

FIGURE 15.9
Polymer resin manufacturing process

be combined, sacrificing detail. The ISA SP-88 style representation described below is more convenient. The vessel is assigned to the procedure which is subdivided into operations that represent charging, blending, and transfer. The illustrative example of SchedulePro that follows provides additional information on modeling, scheduling, and managing of multiproduct batch plants with recipe-based scheduling tools. This example plant manufactures polymer resins.

15.3.2 Polymer Resin Manufacturing Example

Various polymer resins are manufactured using emulsion polymerization processes. One such process is depicted in Figure 15.9. In this process, water, monomers, co-monomers, initiators, and emulsifiers are first charged to the polymerization reactor. After the solids are formed, the emulsifiers are washed out and the solids are centrifuged into totes or bins before being dried and transferred to inventory for loading. Such plants are usually equipped with multiple production lines, manufacturing a large number of resins differing in composition and processing conditions.

Note that the material of this example corresponds to the "Polymer Resin" example of SchedulePro. Its ReadMe file includes the figures of this example in color and it is available online [146]. A functional evaluation version of SchedulePro is available online as well [147].

The Language of SchedulePro

SchedulePro addresses the need to represent repetitive batch manufacturing by utilizing the concept of a recipe. A recipe is a template or description of how to make one batch of something. Using terminology from the batch control ISA SP-88 standard [139], a master batch or recipe for a process consists of one or more unit procedures. A unit procedure ("procedure" for short) is a distinct processing step that utilizes at least one primary piece of equipment for its entire duration. For instance, the unit procedures of the process depicted in Figure 15.9 include Polymerization, Washing, Centrifugation, Storage, and Drying. Unit procedures are further divided into operations. Operations describe distinct sub-steps in a unit procedure. Operations may require other resources such as labor, materials, utilities, auxiliary equipment, and staff.

The relative timing of the various operations is determined by the operation's duration and by scheduling relationships among operations. An operation's duration may be fixed, rate-based (i.e., dependent on the amount of material processed), inventory-dependent (i.e., related to the time it takes for a storage unit to reach a specified level), or specified to be equivalent to the duration of another operation. Each operation's timing may be set relative to the start of a batch or the start or end time of another operation. Flexible shifts related to operation start times can also be specified, in order to allow an operation to be delayed automatically if some of its required resources are not available at its scheduled start time [148].

Once the necessary recipes have been defined, a schedule can be generated. The schedule represents a specific plan that defines which recipes are executed when and with what resources. The schedule is created based on planned production orders or campaigns. A campaign represents a request for a certain amount of product or for a series of batches of a particular recipe. Each batch represents the execution of a single recipe at a specific time and with specific resources. Each campaign has a release date representing its earliest start time and several scheduling options such as a due date and a start time relative to the start or end of some other campaign. In addition, the cycle time for a multi-batch campaign (i.e., the time interval between the start of consecutive batches) can either be fixed by the user or it can be set to an (estimated) minimum value plus some user-defined slack. Campaigns may also contain pre-production and post-production activities to account for time spent setting up or cleaning out equipment. These non-production activities are treated as regular operations that may utilize materials and other resources.

Like other finite capacity scheduling tools, SchedulePro schedules production of campaigns while respecting constraints stemming from resource unavailability (e.g., facility or equipment outages) or availability limitations (e.g., equipment can only be used by one procedure at a time). Conflicts (i.e., violations of constraints) can be resolved by exploiting alternative resources declared as candidates in pools, introducing delays or breaks if this flexibility has been declared in the corresponding operations, or moving the start of a batch to a timeframe where the required resources are available. The automatically generated schedule can subsequently be modified by the user. Through a mix of automated and manual scheduling, users can formulate a production plan that is feasible and satisfies their production objectives. Table 15.4 displays the procedures, their operations, and the operation durations of the base-case polymer resin recipe. Each procedure requires a piece of equipment. In many cases, there may be several alternative equipment units available for a procedure. This is shown by the multiple equipment items (enclosed in brackets) associated with certain procedures in Table 15.4. For instance, polymerization can be carried out in reactor R-1 or R-2. Note that there is a single transfer line (RMTL) for charging materials into the two reactors, and only one reactor can be charged at a time. In the recipe specification, the transfer line (RMTL) is associated with the "Charge Materials" operation (as auxiliary equipment) of the Polymerization procedure because its occupancy is limited to that activity.

The information on operation durations and scheduling links is translated by the tool into a Gantt chart that enables users to visualize a batch process. Figure 15.10 displays the Gantt chart of the recipe described above. The bar at the top repre-

TABLE 15.4

Recipe structure for the polymer resin process

Procedure	Main Equipment	Operations	Duration
Polymerization	Reactors [R-1, R-2]	Charge Materials [RMTL] React Transfer to Wash Vessel Clean Reactor	2 h 16 h 3 h 4 h
Washing	Wash Vessels [WV-1, WV-2]	Receive Crude Wash Feed Centrifuge	3 h 28 h 3.5 h
Centrifugation	Centrifuge [CF-1]	Centrifuge	3.5 h
Storage	Totes [ST]	Receive from Centrifuge Transfer to Dryer	3.5 h 1 h
Drying	Dryer [DR-1]	Load Dryer Dry Unload	1 h 9 h 1 h

sents the duration of the whole recipe, whereas the dark and gray bars underneath represent the duration of each procedure and its associated operations, respectively. Each bar of Figure 15.10 has a label to the right of it to identify the specific operation/procedure that it represents and its duration is displayed in parentheses. The grid on the left-hand-side of the chart displays the same information in tabular form.

In addition, SchedulePro provides a view of a recipe in block-diagram format (Figure 15.11). Each rectangle in this diagram represents a procedure. The gray arrows that connect the various operations of the procedures indicate scheduling links among those operations. If the recipe includes information on consumption and generation of materials, those are indicated by black horizontal arrows linked to operations. The recipe Gantt chart and block diagram facilitate editing in addition to visualization of a batch process.

SchedulePro provides a number of other recipe views for easy editing of labor, auxiliary equipment, materials, and other resources.

If a batch process is already modeled in SuperPro Designer, its recipe can be exported to SchedulePro via a recipe database. The same is possible with batch process automation and manufacturing execution systems that follow the ISA S-88 standards for batch recipe representation. Alternatively, recipes can be created by users directly in SchedulePro.

Figure 15.12 displays the equipment occupancy chart (EOC) of a 6-batch campaign of this simple process. The different colors correspond to the six different batches (AA-1 through AA-6). Equipment is displayed on the y-axis and time on the x-axis. The bars on the top line represent the occupancy of the raw material transfer line (RMTL). The next two lines represent the occupancy of the two reactors (R-1 and R-2). Notice that each reactor is used only three times in this campaign. This is because there are two reactors available, and the six batches alternate between these two reactors.

The next two lines represent the occupancy of the two wash vessels (WV-1 and

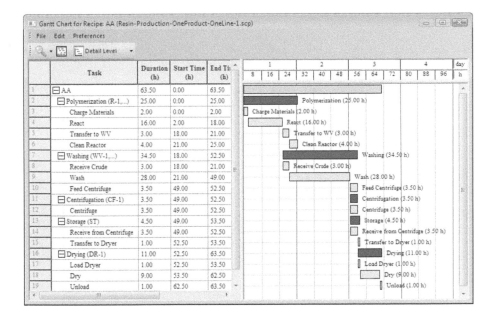

FIGURE 15.10
Gantt chart of the polymer resin recipe

WV-2). Like the reactors, each wash vessel is used three times in the campaign. These vessels are the current cycle time bottleneck of the process and they determine the minimum time between consecutive batches. This can be visualized by the fact that there is no idle time between the consecutive batches of the wash vessels.

Note that the schedule of Figure 15.12 was generated using the "Automatic" scheduling mode of SchedulePro that attempts to schedule a multi-batch campaign based on a fixed cycle time (the minimum by default). Solutions based on other scheduling modes (e.g., the ASAP mode) are displayed in the sections that follow.

Scheduling a Multiline and Multiproduct Plant

For the sake of simplicity, let us assume that only three different polymer resins are to be produced (AA, BB, and CC). Resin AA corresponds to the base case recipe described in the previous section. The recipes of resins BB and CC include exactly the same processing steps but have different durations for key operations, as shown in Table 15.5.

Furthermore, these resins will be produced in a more-complex plant than the 2-reactor production scenario described in the previous section. Figure 15.13 displays the equipment layout of the plant, which includes a single raw material transfer line (RMTL), four polymerization reactors, six wash vessels, two centrifuges, a large number of storage totes, and two dryers. Furthermore, the reactors, wash vessels, and centrifuges are split into two production suites (lines). Suite 1 includes two reactors (R-1 and R-2), three wash vessels (WV-1, WV-2, and WV-3), and one centrifuge

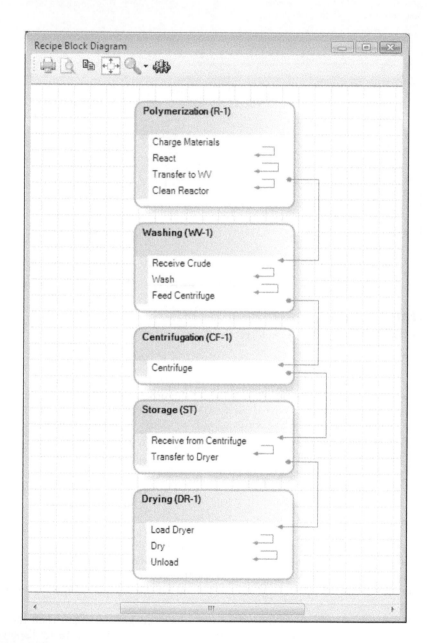

FIGURE 15.11

Block diagram of the polymer resin recipe

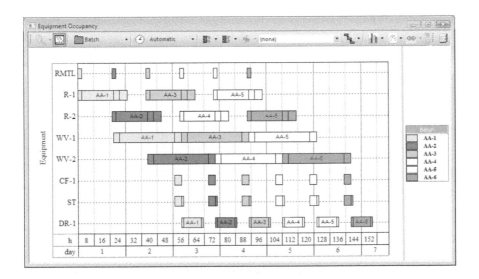

FIGURE 15.12
Equipment occupancy chart for six batches of the same resin

TABLE 15.5
Key operations of the three polymer resins

Product	Reaction Time (h)	Wash Time (h)	Centrifugation Time (h)	Drying Time (h)
AA	16	28	3.5	9
BB	14	36	3.0	14
CC	14	28	4.0	12

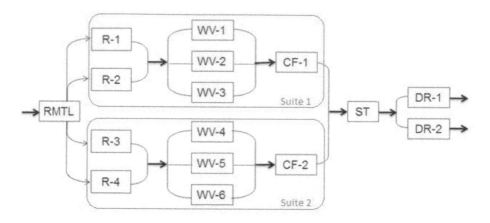

FIGURE 15.13
Resin manufacturing plant equipment layout

(CF-1). Suite 2 also includes two reactors (R-3 and R-4), three wash vessels (WV-4, WV-5, and WV-6), and one centrifuge (CF-2). A batch that is started in a certain suite must remain in that suite until it goes through centrifugation. The centrifuged solids (coming out of CF-1 and CF-2) are then stored in mobile totes (ST). The two dryers (DR-1 and DR-2) are in another part of the building, and are not linked to the production suites. Therefore any batch can go to either dryer. Furthermore, all reactors have the same size, all wash vessels have the same size, and both centrifuges have the same throughput capacity. As a result, any polymer resin recipe can be run in either suite, and campaigns of the different recipes will need to share equipment (i.e., there isn't a dedicated equipment train associated with each specific recipe). As before, the raw materials are fed through a single transfer line (RMTL), and only one reactor can be charged at a time.

Additional cleaning is required whenever product changes occur in a reactor. The duration of cleaning is product sequence dependent according to the changeover matrix of Figure 15.14. For instance, a switch from AA to BB or CC requires additional cleaning of 4 h whereas a switch from BB or CC to any other recipe requires a longer cleaning of 6 h. Also, if a reactor remains idle for more than 48 h, it requires a pre-cleaning of 4 h before it can be used.

The relative demand for the three products is: $AA/BB/CC = 2/4/1$. Consequently, the plant tends to schedule two batches of AA, followed by four batches of BB, followed by one batch of CC in a repeating pattern. Figure 15.15 displays a typical production schedule in which the 2/4/1 pattern for AA/BB/CC is repeated three times. Each product is displayed with a specific color (a color version of this chart is available online [146]). The white bars represent changeover activities whenever a product switch occurs in a reactor. Furthermore, the longer procedure bars display their specific batch identifiers (e.g., the bars that are labeled with "AA-1-2" correspond to batch #2 of campaign #1 of product AA).

The schedule of Figure 15.15 corresponds to the ASAP (as soon as possible) scheduling mode of SchedulePro. In this mode, each new batch is scheduled to begin

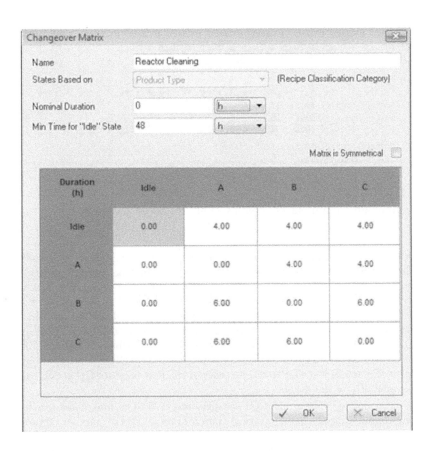

FIGURE 15.14
Reactor cleaning changeover matrix

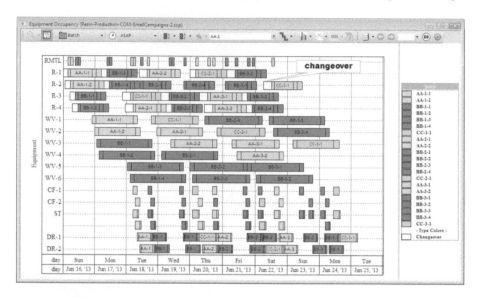

FIGURE 15.15

Multi-campaign equipment occupancy chart for three products

as soon as the equipment required to process it becomes available. Notice that the start times of the first two batches of AA are staggered because the single raw material transfer line (RMTL) can only be used to supply one reactor at a time. In contrast, the start of the first two batches of product BB is determined by the dryers. Note that it is possible to start those two batches a little earlier and hold the material in the storage totes until the dryers become available. That mode of operation can be modeled in SchedulePro by adding flexible start time shifts to appropriate operations. The start of the next two batches of BB is determined by the occupancy of reactors R-1 and R-2. The start of the first batch of CC is determined by the availability of the dryers.

The makespan of this production scenario is 221 h. The makespan can be reduced by running larger campaigns and therefore reducing the number of product changeovers. Figure 15.16 represents such a solution. In this mode, the makespan of the schedule is reduced from 221 to 208 h (6% reduction), which is equivalent to a throughput increase of 6%. However, a plant operating in this mode will need to maintain larger inventories of products in order to meet the same product demand. This well-known tradeoff between just-in-time manufacturing (i.e., running many small campaigns) and plant throughput (running fewer but larger campaigns) can be readily evaluated with simulation and scheduling tools.

Considering Shared Resources

In addition to equipment unavailability, other resource constraints can have a substantial impact on the actual capacity of a batch manufacturing plant. These resources can include labor, utilities, raw materials, etc. If insufficient resources are

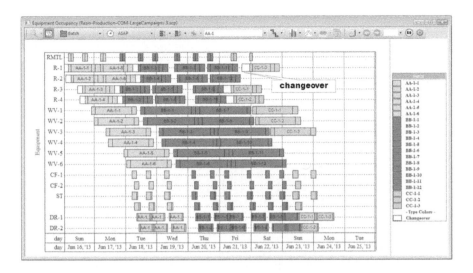

FIGURE 15.16
Equipment occupancy chart with longer campaigns

available at a given time, processing of a batch may be delayed, leading to decreased plant efficiency and throughput, as well as late orders. To demonstrate evaluation of a resource, labor requirements were added to various operations of the recipe shown in Figure 15.10. For instance, the Charge Materials operation in the Polymerization procedure was assigned one operator. Other operations, such as the React operation in the Polymerization procedure, were assigned a partial operator. In other words, if an operation only requires 1/10th of an operator's time (since it only requires occasional monitoring and intervention), a value of 0.1 operator was assigned to it. Figure 15.17 shows the labor demand associated with the production schedule of Figure 15.15.

Notice that there are large swings in labor demand over the course of a week, which means the utilization rate of the labor force is somewhat inefficient. As can be seen, the labor demand for this production schedule peaks on day 5 (Thursday) at a maximum of 5.7 operators. In other words, 5.7 operators (or more realistically, 6 operators) would be required in order to perform the activities that are scheduled for this particular time. If fewer than 6 operators are available, certain operations will need to be delayed. In some cases, this may delay a batch and reduce plant throughput. In other cases, batch completion will not be delayed because the operation shift does not impact an activity on the critical path. SchedulePro has the ability to generate schedules that respect labor constraints which may vary with work shifts and weekends. Other constraints that may optionally be considered include the limited availability of utilities and materials.

Production Tracking and Rescheduling

Real-world scheduling of multiproduct batch plants involves a repeating cycle of the following activities:

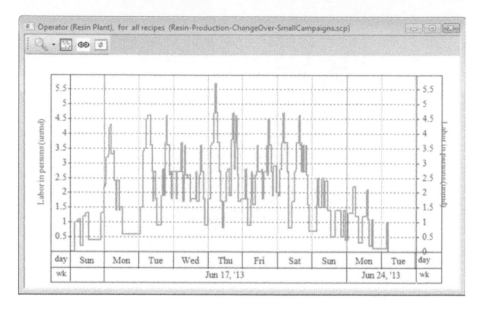

FIGURE 15.17

Labor demand chart

1. Adding new batches to the current schedule
2. Updating the schedule to account for actual process information
3. Adjusting for unforeseen events

Many plant schedulers use spreadsheets to aid with these tasks, but spreadsheets can be difficult to update and maintain, and they cannot comprehensively account for complex dependencies. Recipe-based scheduling tools greatly reduce the effort required for managing plants on a day-to-day basis.

The completion of certain manufacturing tasks is often delayed due to variability in operator skills, performance of process controllers, equipment failures, etc. Such delays may lead to scheduling conflicts with future activities. For example, let's assume that we have begun running the production schedule displayed in Figure 15.15. We are approximately 1.5 days into our schedule. This is shown in the equipment occupancy chart in Figure 15.18, where the vertical line in the middle of the day on Monday, June 17th represents the current time. The current time line results in the division of activities into three categories: completed (displayed by a crossed hatch pattern), in-progress (diagonal hatch), and not-started (filled pattern).

Let's assume that the polymerization reaction in batch BB-1-3 (the third batch of the first BB campaign that utilizes R-1) took 4 h longer than planned. As a result, the end of batch BB-1-3 in reactor R-1 has a conflict with the scheduled start time of batch AA-2-2 that was scheduled to utilize the same reactor. This conflict is displayed in Figure 15.18 with a new line on the equipment occupancy chart directly below line R-1 (this new line is labeled with an exclamation point and the conflicting procedure associated with batch AA-2-2 is outlined in red). Furthermore, the delay in the polymerization operation in batch BB-1-3 has delayed subsequent operations

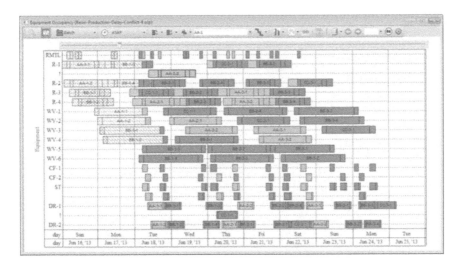

FIGURE 15.18

Tracking the production schedule

in BB-1-3 and created a conflict with batch CC-1-1 for the use of dryer DR-1. In other words, a delay in a single operation (the BB-1-3 reaction) has already created conflicts with two other batches in two different equipment units.

At this point, the next logical step is to resolve the conflicts created by the initial delay. To resolve these conflicts, the scheduler could delay each individual procedure which has a conflict (e.g., delay the Polymerization procedure for batch AA-2-2, and delay the Drying procedure for batch CC-1-1). Note that this may create delays in other procedures which are scheduled later. For instance, if the Drying procedure for batch CC-1-1 is delayed, it will create a conflict with the Drying procedure for batch AA-2-2. As a result, conflict resolution may need to be performed iteratively, with the earliest conflicts being addressed first. Alternatively, all conflicts could be resolved automatically by the tool by rescheduling the affected batches and treating them with a lower priority than the batches that have already started. Figure 15.19 shows the updated production schedule after all conflicts have been resolved in this manner.

Contemporary scheduling tools are equipped with databases (typically SQL Server or Oracle) for tracking the status of production as a function of time, communicating the data to various stakeholders, and archiving completed batches and campaigns. After archiving a campaign, it can be deleted from the production schedule. This allows the scheduler to focus on the current and future campaigns.

In addition to storing historical data and tracking the status of production, central databases facilitate communication with enterprise resource planning (ERP) and related tools. For instance, an ERP tool may deposit an unscheduled campaign in the database (representing a new work order). Such a campaign can be imported into the scheduling tool, scheduled, and executed. The status of the campaign along with information on consumption and generation of materials can be communicated back to the ERP tool from time to time.

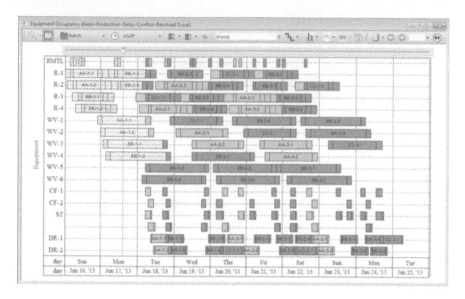

FIGURE 15.19
The updated production schedule

A variety of third-party reporting tools are available for viewing data stored in SQL Server and Oracle databases. Users can create their own reports which can be viewed through Internet browsers and smart phones. Thus, project managers can remotely monitor the status of campaigns and projects of their organization on an on-going basis. Simplified reports and metrics that provide high level information are recommended for the executives. Detailed reports that focus on the activities of a specific production line for a specific date or shift are useful for providing execution instructions to operators and line supervisors.

15.4   Summary

Batch process simulation and production scheduling tools can play an important role throughout the life-cycle of product development and commercialization. In process development, process simulation tools are becoming increasingly useful as a means to analyze, communicate, and document process changes. During the transition from development to manufacturing, they facilitate technology transfer and process fitting. Production scheduling tools play a valuable role in manufacturing as well. They are used to generate production schedules based on the accurate estimation of plant capacity and resource constraints, thus minimizing late orders and reducing inventories. Such tools also facilitate capacity analysis and debottlenecking tasks. Finally, they are useful for performing ongoing tracking and updating of the manufacturing schedule.

The batch process industries have begun making significant use of process simulation and scheduling tools. Increasingly, universities are incorporating the use of such tools in their curricula. In the future, we can expect to see increased use of these technologies and tighter integration with other enabling IT technologies, such as supply chain tools, manufacturing execution systems (MES), batch process control systems, process analytical technology (PAT), etc. The result will be more robust and efficient manufacturing processes.

Bibliography

[1] Ki-Joo Kim. *Solvent Selection and Recycling: A Multiobjective Optimization Framework for Separation Processes*. PhD thesis, Carnegie Mellon University, Department of Environmental Engineering, Pittsburgh, PA, August 2001.

[2] S. Branauer, P. Emmett, and E. Teller. On a theory of the van der waals adsorption of gases. *J. Am. Chem. Soc*, 62:1723–1732, 1940.

[3] J. F. Richardson, J. H. Harker, and J. R. Backhurst. *Coulson and Richardsons CHEMICAL ENGINEERING*. Butterworth-Heinemann, London, 5th edition, 2002.

[4] J. Ward, D. Mellichamp, and M. Doherty. Choosing an operating policy for seeded batch crystallization. *AIChE J.*, 52(6):2046–2054, 2006.

[5] R. L. Earle and M.D. Earle. *Unit Operations in Food Procesing*. NZFIST(inc), http://www.nzifst.org.nz/unitoperations, the web edition, 1983.

[6] R. Harrison, P. Todd, S. Rudge, and D. Petrides. *Bioseparations Science and Engineering*. Oxford University Press, Oxford, 2003.

[7] U. Diwekar. *Batch Distillation: Simulation, Optimal Design, and Control*. CRC Press, Boca Raton, 2nd edition, 2012.

[8] G. P. Distefano. Stability of numerical integration techniques. *AIChE J*, 14:946–95, 1968.

[9] P. Henrici. *Discrete variable methods in ODEs*. Prentice-Hall Inc.,, John Wiley & Sons, New York, 1962.

[10] G. H. Golub and J. M. Ortega. *Scientific Computing and Differential Equations*. Academic Press, San Diego, 1992.

[11] C. F. Curtis and J. O. Hirschfelder. Integration of stiff equations. *National Academy of Sciences*, 38:235–243, 1952.

[12] C. W. Gear. *Numerical initial value problems in ordinary differential equations*. Prentice-Hall Inc.,, Englewood Cliffs, N. J., 1971.

[13] J. Lambert. *Computational methods in ordinary differential equations*. Prentice-Hall Inc., Wiley, New York, 1973.

[14] J. C. Butcher. *The numerical analysis of ordinary differential equations*. John Wiley & Sons, New York, 1987.

[15] S. O. Fatunla. *Numerical methods for initial value problems in ordinary differential equations*. Academic Press, San Diego, 1988.

[16] B. Galerkin. Rods and plates. series in some problems of elastic equilibrium of rods and plates. *JVestn. Inzh. Tech. (USSR)*, 19:897, 1915.

[17] R. Frazer, W. Jones, and S. Skan. Approximations to functions and to the solutions of differential equations, great britain aerospace research council london. report and memo no. 1799. *Great Britain Air Ministry Aerospace Research Communications Technical Report*, 1:735, 1937.

[18] C. Lanczos. Trigonometric interpolation of empirical and analytic functions. *Journal of Mathematical Physics*, 17:123, 1938.

[19] J. Villadsen. *Selected Approximation Methods for Chemical Engineering Problems*. Instituttet for Kemiteknik, Lyngby, Denmark, 1970.

[20] J. V. Villadsen and W. E. Stewart. Solution of boundary-value problems by orthogonal collocation. *Chem. Eng. Sci*, 22:1483, 1967.

[21] B. Finlayson. *The Method of Weighted Residuals and Variational Principles*. Academic Press, New York, 1972.

[22] J. Villadsen and M. L. Michalesen. *Solution of Differential Equation Models by Polynomial Approximation*. Prentice-Hall, Englewood Cliffs, NJ, 1978.

[23] W. Stewart, K. Levien, and M. Morari. Simulation of fractionation by orthogonal collocation. *Chem. Eng. Sci*, 40:409, 1985.

[24] O. Levenspiel. *Chemical Reaction Engineering*. John Wiley & Sons, New York, 3rd edition, 1999.

[25] K. Denbigh. Optimal temperature sequences in reactors. *Chem. Eng. Sci*, 8:125, 1958.

[26] B. Srinivasan, S. Palanki, and D. Bonvin. Dynamic optimization of batch processes i. characterization of the nominal solution. *Comp. Chem. Eng*, 27:1, 2003.

[27] J. Villadsen. Batch reactors in the bioindustries. *in Batch Processes, Korovessi and Linninger editors*, CRC Press:83, 2005.

[28] J. Monod. The growth of bacterial cultures. *Annual Review of Microbiology*, 3:371, 1949.

[29] L. Menten and M. I. Michaelis. Die kinetik der invertinwirkung. *Biochem*, 49:333, 1913.

[30] E. Herrera, G. Hernandez, A. Moran, M. Flores, J. Cordova, R. Femat, and D. Diaz-Montano. Modelado matematico del cultivo en lote y continuo del proceso fermentativo del tequila. *Congreso Anual 2009 de la Asociacion de Mexico de Control Automatico. Zacatecas, Mexico*, 2009.

[31] H. Noureddini and D. Zhu. Kinetic of transesterification of soybean oil. *J. Am. Oil. Chem. Soc*, 74:1457, 1997.

[32] P. Benavides and U. Diwekar. Optimal control of biodiesel production in a batch reactor part i: Deterministic control. *Fuel*, 94:211, 2012.

[33] Y. Dennis, W. Xuan, and M. Leung . A review on biodiesel production using catalyzed transesterification. *Applied Energy*, 83:1083, 2010.

[34] P. Benavides and U. Diwekar. Studying various optimal control problems in biodiesel production in a batch reactor under uncertainty. *Fuel*, 103:585, 2013.

[35] L. Rayleigh. On distillation of binary mixtures. *Philosophical Magazine (vi)*, 4:521, 1902.

[36] Urmila M. Diwekar. *Batch Distillation: Simulation, Optimal Design and Control*. Taylor and Francis, Washington D.C., 1996.

[37] Arthur G. Davidyan, Valerii N. Kiva, George A. Meski, and Manfred Morari. Batch distillation column with a middle vessel. *Chem. Eng. Sci.*, 49(18):3033–3051, 1994.

[38] Sigurd Skogestad, Bernd Wittgens, Rajab Litto, and Eva Sørensen. Multivessel batch distillation. *AIChE J.*, 43(4):971–978, 1997.

[39] W. L. McCabe and E. W. Thiele. *Ind. Eng. Chem.*, 17:605–611, 1925.

[40] E. H. Smoker and A. Rose. Graphical determination of batch distillation curves for binary mixtures. *Transactions of American Institute of Chemical Engineers*, 36:285–293, 1940.

[41] M. J. P. Bogart. The design of equipment for fractional batch distillation. *Trans. AIChE*, 33:139, 1937.

[42] I. Coward. The time-optimal problems in binary batch distillation. *Chem. Eng. Sci.*, 22:503–516, 1966.

[43] G. P. Distefano. Mathematical modeling and numerical integration of multicomponent batch distillation equations. *AIChE J.*, 14:190–199, 1968.

[44] Urmila M. Diwekar. Understanding batch distillation process principles with multibatchds. *Computer Applications in Chemical Engineering Education*, 4(4):275–284, 1996.

[45] Urmila M. Diwekar and K. P. Madhavan. Batch-dist: A comprehensive package for simulation, design, optimization and optimal control of multicomponent, multifraction batch distillation columns. *Comp. Chem. Eng.*, 15(12):833–842, 1991.

[46] Urmila M. Diwekar, K. P. Madhavan, and R. W. Swaney. Optimization of multicomponent batch distillation column. *Ind. Eng. Chem. Res.*, 28:1011–1017, 1989.

[47] A. O. Converse and G. D. Gross. Optimal distillate rate policy in batch distillation. *Ind. Eng. Chem. Fundamentals*, 2:217–221, 1963.

[48] Urmila M. Diwekar, R.K. Malik, and K.P. Madhavan. Optimal reflux rate policy determination for multicomponent batch distillation columns. *Comp. Chem. Eng.*, 11:629–637, 1987.

[49] Jeffrey S. Logsdon, Urmila M. Diwekar, and Lorenz T. Biegler. On the simultaneous optimal design and operation of batch distillation columns. *Transactions of IChemE*, 68:434–444, 1990.

[50] S. Farhat, M. Czernicki, L. Pibouleau, and S. Domenech. Optimization of multiple-fraction batch distillation by nonlinear programming. *AIChE J.*, 36:1349–1360, 1990.

[51] Jeffrey S. Logsdon and Lorenzo T. Biegler. Accurate determination of optimal reflux policies for the maximum distillate problem in batch distillation. *Ind. Eng. Chem. Res.*, 32:692–700, 1993.

[52] Urmila M. Diwekar. Unified approach to solving optimal design - control problems in batch distillation. *AIChE J.*, 38(10):1551–1563, 1992.

[53] L. H. Kerkhof and H. J. M. Vissers. On the profit of optimal control in batch distillation. *Chem. Eng. Sci.*, 33:961–970, 1978.

[54] Pu Li, G. Wozny, and E. Reuter. Optimization of multiple-fraction batch distillation with detailed dynamic process model. *Institution of Chemical Engineers Symposium Series*, 142:289–300, 1997.

[55] Iqbal M. Mujtaba and Sandro Macchietto. Efficient optimization of batch distillation with chemical reaction using polynomial curve fitting technique. *Industrial Engineering Chemistry Proceedings Design and Development*, 36:2287–2295, 1997.

[56] H. I. Furlonge, C. C. Pantelides, and Eva Sørensen. Optimal operation of multivessel batch distillation columns. *AIChE J.*, 45(4):781–801, 1999.

[57] Shinji Hasebe, Masaru Noda, and Iori Hashimoto . Optimal operation policy for total reflux and multi-batch distillation systems. *Comp. Chem. Eng.*, 23:523–532, 1999.

[58] Urmila M. Diwekar. An efficient design method for binary azeotropic batch distillation. *AIChE J.*, 37:1571–1578, 1991.

[59] Jayant R. Kalagnanam and Urmila M. Diwekar. An application of qualitative analysis of ordinary differential equations to azeotropic batch distillation. *Artificial Intelligence in Engineering*, 8:23–32, 1993.

[60] Christine Bernot, Michael F. Doherty, and Michael F. Malone. Patterns of composition change in multicomponent batch distillation. *Chem. Eng. Sci.*, 45:1207–1221, 1990.

[61] Christine Bernot, Michael F. Doherty, and Michael F. Malone. Feasibility and separation sequence in multicomponent batch distillation. *Chem. Eng. Sci.*, 46:1311–1326, 1991.

[62] Weiyang Cheong and Paul I. Barton. Azeotropic distillation in a middle vessel batch column. 1. model formulation and linear separation boundaries. *Ind. Eng. Chem. Res.*, 38:1504–1530, 1999.

[63] P. E. Cuille and G. V. Reklaitis. Dynamic simulation of multicomponent batch rectification with chemical reactions. *Comp. Chem. Eng.*, 10(4):389–398, 1986.

[64] R. M. Wajge, J. M. Wilson, J. F. Pekny, and G. V. Reklaitis. Investigation of numerical solution approaches to multicomponent batch distillation in packed beds. *Ind. Eng. Chem. Res.*, 36:1738–1746, 1997.

[65] D. M. Hitch and R. W. Rousseau. Simulation of continuous contact separation processes multicomponent batch distillation. *Ind. Eng. Chem. Res.*, 27:1466–1473, 1988.

[66] R. M. Wajge and G. V. Reklaitis. An optimal campaign structure for multicomponent batch distillation with reversible reaction. *Ind. Eng. Chem. Res.*, 37:1910–1916, 1998.

[67] Sandro Macchietto and Iqbal M. Mujtaba. Design of operation policies for batch distillation. In G. V. Reklaitis, editor, *NATO Advanced Study Institute Series F143*, pages 174–215. Springer-Verlag, Berlin, 1996.

[68] L. U. Kreul, A. Górak, and P. I. Barton. Dynamic rate-based model for multicomponent batch distillation. *AIChE J.*, 45(9):1953–1961, 1999.

[69] Joseph F. Boston, Herbert I Britt, Siri Jirapongphan, and V. B. Shah. An advanced system for the simulation of batch distillation operations. In Richard S.H. Mah and Warren D. Seider, editors, *Foundations of Computer-Aided Chemical Process Design*, volume 2, pages 203–237. Engineering Foundation, New York, 1983.

[70] Urmila M. Diwekar. *Introduction to Applied Optimization*. Springer, New York, NY, second edition, 2008.

[71] M. W. Carter and C. C. Price. *Operations Research: A Practical Introduction*. CRC Press, New York, 2001.

[72] H. A. Taha. *Operations Research: An Introduction*. Prentice-Hall, Upper Saddle River, NJ, sixth edition, 1997.

[73] W. L. Winston. *Operations Research: Applications and Algorithms*. PWS-KENT, Boston, second edition, 1991.

[74] N. Karush. PhD thesis, Department of Mathematics, University of Chicago, 1939.

[75] H. W. Kuhn and A. W. Tucker. Nonlinear programming. In Neyman, editor, *Proceedings of Second Berkeley Symposium on Mathematical Statistics and Probability*, University of California Press, Berkeley, 1951.

[76] A. M. Geoffrion. Generalized benders decomposition. *Journal of Optimization Theory and Applications*, 10:237–260, 1972.

[77] M. A. Duran and Ignacio E. Grossmann. An outer-approximation algorithm for a class of mixed-integer nonlinear programs. *Mathematical Programming*, 36:307–339, 1986.

[78] N.E. Collins, R. W. Eglese, , and B. L. Golden. Simulated annealing–an annotated biography. *American Journal of Mathematical and Management Science*, 8(3):209, 1988.

[79] P. J. M. VanLaarhoven and E. H. Aarts. *Simulated Annealing Theory and Applications*. D. Riedel, Holland, 1987.

[80] D.E. Goldberg. *Genetic Algorithms in Search, Optimization, and Machine Learning*. Addison-Wesley, Reading MA, 1989.

[81] V. G. Boltyanskii, R. V. Gamkrelidze, and L. S. Pontryagin. On the theory of optimum processes (in russian). *Doklady Akad. Nauk SSSR*, 110:1, 1956.

[82] L. S. Pontryagin. Some mathematical problems arising in connection with the theory of automatic control system (in russian). In *Session of the Academic Sciences of the USSR on Scientific Problems of Automatic Industry*, Paris, France, October 1956.

[83] Richard Bellman. *Dynamic Programming*. Princeton University Press, Princeton, New Jersey, 1957.

[84] R. Aris. *The Optimal Design of Chemical Reactors*. Academic Press, London, 1961.

[85] Douglas N. Dean, Michael J. Fuchs, John M. Schaffer, and Ruben G. Carbonell. Batch absorption of co_2 by free and microencapsulated carbonic anhydrase. *Ind. Eng. Chem., Fundam.*, 16(4):452, 1977.

[86] W. McCabe, J. Smith, and P. Harriott. *Unit Operations of Chemical Engineering*. Mc Graw Hill, New York, 7 edition, 2005.

[87] P. Belter, E. Cussler, and W. Hu. *Bioseparations*. Wiley-Interscience, 1988.

[88] S. Branauer, P. Emmett, and E. Teller. Adsorption of gases in multimolecular layers. *J. Am. Chem. Soc*, 60:309–319, 1938.

[89] P. Emmett and T. De Witt. Determination of surface areas. *Ind. Eng. Chem. (Anal.)*, 13:28–35, 1941.

[90] W. Harkins and G. Jura. An adsorption method for the determination of the area of a solid without the assumption of a molecular area and the area occupied by nitrogen molecules on the surface of solids. j. amer. chem. soc. 66 (1944) 1366. surface of solids. part xiii. *J. Chem. Phys.*, 11:431–442, 1943.

[91] I. Langmuir. The adsorption of gases on plane surfaces of glass, mica and platinum. *J. Am. Chem. Soc*, 40:1361–1370, 1918.

[92] C. Sheindorf and M. Rebhun. A freundlich-type multicomponent isotherm. *Journal of Colloid and Interface Science*, 79(1):136–142, 1981.

[93] A. Martin and Synge R. . A new form of chromatogram employing two liquid phases. i. theory of chromatography. *Biochem J.*, 35:1358, 1941.

[94] J. Van Deemter, F. Zuiderweg , and A. Klinkenberg A. Logitudinal diffusion and resistance to mass transfer of non-ideality in chromatography. *Chem. Eng. Sci.*, 5:271, 1956.

[95] P. Hill. Batch crystallization. In E. L Korovessi and A. A.Linninger, editors, *Batch Processes*. CRC Press: Taylor & Francis, 2006.

[96] A. Randolph and M. Larson. *Theory of Particulate Processes*. Academic Press, San Diego, 1988.

[97] J. Mullin. *Crystallization*. Butterworth-Heinemann, London, 3rd edition, 1993.

[98] K. Yenkie and U. Diwekar. Stochastic optimal control of seeded batch crystallizer applying the ito process. *AIChE J.*, 51(11):3000–3006, 2005.

[99] Q. Hu, S. Rohani, and A. Jutan. Modelling and optimization of seeded batch crystallizers. *Computers and Chemical Engg*, 29:911–918, 2005.

[100] D. Shi, N. El-Farra, M. Li, P. Mhaskar, and P. Christofides. Predictive control of particle size distribution in particulate processes. *Chem. Eng. Sci.*, 61:268–280, 2006.

[101] S. Kumar and D. Ramakrishna. On the solution of population balance equations by discretization -iii. nucleation, growth and aggregation of particles. *Chem.Eng. Sci.*, 52(24):4659–4679, 1997.

[102] Q. Hu, S. Rohani, and A. Jutan. New numerical method for solving the dynamic population balance equations. *AIChE J.*, 51(11):3000–3006, 2005.

[103] J. Corriou and S. Rohani. A new look at optimal control of a batch crystallizer. *AIChE J.*, 54(12):3188–32060, 2008.

[104] A. Flood. Thoughts on recovering particle size distributions from the moment form of the population balance. *Dev. Chem. Eng. Mineral Process*, 10(5/6):501–519, 2002.

[105] V. Johna, I. Angelovb, A.A. Onculc, and D. Theven. Techniques for the reconstruction of a distribution from a finite number of its moments. *Chem.Eng. Sci.*, 62:2890 2904, 2007.

[106] A. Jones. Optimal operation of a batch cooling crystallizer. *Chem Eng. Sci.*, 29:1075–1087, 1974.

[107] J. Rawlings, S. Miller, and W. Witkowski. Model identification and control of solution crystallization processes: A review. *Ind Eng Chem. Res.*, 32:1275–1296, 1993.

[108] C. Chang and M. Epstein. Identification of batch crystallization control strategies using characteristic curves. In Epstein, editor, *Nucleation, Growth and Impurity Effects in Crystallization Process Engineering*, New York, 1982. AIChE.

[109] D. Ma and R. Braatz. Robust identification and control of batch processes. *Comput Chem Eng.*, 27:1175–1184, 2003.

[110] S. Chung, D. Ma, and R. Braatz. Optimal seeding in batch crystallization. *Chem Eng. Sci.*, 77:590–596, 1999.

[111] J. Eaton and J. Rawlings. Feedback control of chemical processes using on-line optimization techniques. *Comput Chem Eng.*, 14:469–479, 1990.

[112] S. Miller and J. Rawlings. Model identification and control strategies for batch cooling crystallizers. *AIChE J.*, 40:1312–1327, 1994.

[113] L. Feng and K. Berglund. Atr-ftir for determining optimal cooling curves for batch crystallization of succinic acid. *Cryst Growth Des.*, 2:449–452, 2002.

[114] M. Ge, Q. Wang, M. Chiu, T. Lee, C. Hang, and K. Teo. An effective technique for batch process optimization with application to crystallization. *Trans IChemE.*, 78:99 –106, 2000.

[115] G. Zhang and S. Rohani. On-line optimal control of a seeded batch cooling crystallizer. *Chem Eng Sci.*, 58:1887–1896, 2003.

[116] J. Rawlings, C. Sink, and S. Miller. Control of crystallization processes. In *Handbook of Industrial Crystallization*, Boston, 2002. Butterworth Heinemann.

[117] K. Choong and R. Smith. Optimization of batch cooling crystallization. *Chcm Eng Sci.*, 59:313–327, 2004.

[118] H. Suttle . Development of industrial filtration. *The Chemical Engineer*, 314:675, Oct. 1976.

[119] A. Michaels . New separation technique for the cpi. *Chem. Eng. Prog.*, 64:31, 1968.

[120] G. Belfort, R. Davis, and A. Zydney. The behavior of suspensions and macromolecular solutions in crossflow microfiltration. *J. Membrane Sci.*, 96:1, 1994.

[121] A. Zydney and C. Colton . A concentration polarization model for filtrate flux in cross-flow microfiltration of particulate suspension. *Chem. Eng. Commun.*, 47:1, 1986.

[122] R. Perry, D. Green, and J. Maloney eds.. *Section 22, Perry's Chemical Engineer's Handbook*. McGraw-Hill, 7-th edition, 1997.

[123] E. Eckstein, P. Bailey, and A. Shapiro . Self-diffusion of particles in shear flow of a suspension. *J. Fluid Mech.*, 79:191, 1977.

[124] C. A. Mendez, J. Cerda, I. E. Grossmann, I. Harjunkoski, and M. Fahl. State-of-the art review of optimization methods for short-term scheduling of batch processes. *Computers and Chemical Engineering*, 30:913–946, 2006.

[125] H-M. Ku, D. Rajagopalan, and I. Karimi. Scheduling in batch processes. *Chemical Engineering Progress*, 83:35–45, 1987.

[126] G. Applequist, O. Samikoglu, J. Pekny, and G. Reklaitis. Issues in the use, design and evolution of process scheduling and planning systems. *ISA Transactions*, 36(2):81–121, 1997.

[127] C. A. Floudas and X. Lin. Mixed integer linear programming in process scheduling: Modeling, algorithms, and applications. *Annals of Operations Research*, 139(1):131–162, 2005.

[128] M. Pinedo. *Scheduling, Theory, Algorithms and Systems*. Prentice Hall, New Jersey, 2002.

[129] H. Wang. Flexible flow shop scheduling: optimum, heuristics and artificial intelligence solutions. *Expert Systems*, 22:78–84, 2005.

[130] R. Ruiz and C. Maroto. A genetic algorithm for hybrid flowshops with sequence dependent setup times and machine eligibility. *European Journal of Operational Research*, 169:781–800, 2006.

[131] T. Sawik. Mixed integer programming for scheduling flexible flow lines with limited intermediate buffers. *Mathematical and Computer Modeling*, 31:39–52, 2000.

[132] D. Petrides, A. Koulouris, and P. Lagonikos . The role of process simulation in pharmaceutical process development and product commercialization. *Pharmaceutical Engineering*, 22:1, January-February 2002.

[133] F. Hwang . Batch pharmaceutical process design and simulation. *Pharmaceutical Engineering*, page 28, 1997.

[134] A. Toumi, C. Jurgens, C. Jungo, B. Maier, V. Papavasileiou, and D. Petrides. Design and optimization of a large scale biopharmaceutical facility using process simulation and scheduling tools. *Pharmaceutical Engineering*, 30:2, March/April 2002.

[135] G. Plenert and B. Kirchmier . *Finite Capacity Scheduling - Management, Selection, and Implementation, ISBN 0-471-35264-0*. John Wiley & Sons, 2000.

[136] Y. Crama, Y. Pochet, and Y. Wera . A discussion of production planning approaches in the process industry. *Universit Catholique de Louvain, Center for Operations Research and Econometrics, CORE Discussion Papers #2001042*, 2001.

[137] Intelligen. *SuperPro User's Guide, Chapter 2 (Tutorial), The full document can be downloaded in PDF format from http://www.intelligen.com/dem*. 2013.

[138] Intelligen. *Training Videos," http://www.intelligen.com/videos*. 2013.

[139] J. Parshall and L. Lamb. *Applying S88 - Batch Control from a User's Perspective*. ISA - Instrument Society of America, 2000.

[140] Intelligen. *SuperPro Designer SynPharm Example. A functional evaluation version of SuperPro Designer is available online at www.intelligen.com/downloads*. 2013.

[141] D. Petrides, A. Koulouris, and C. Siletti . Throughput analysis and debottle-necking of biomanufacturing facilities, a job for process simulators. *BioPharm*, August, 2002.

[142] M. Peters, , and K. Timmerhaus . *Plant Design and Economics for Chemical Engineers*. McGraw-Hill, New York, 4-th edition, 1991.

[143] J. Mun . *Applied Risk Analysis - Moving Beyond Uncertainty in Business*. John Wiley, Hoboken, New Jersey, 4th edition, 2004.

[144] E.C. Achilleos, J.C. Calandranis, and D.P. Petrides. Quantifying the impact of uncertain parameters in the batch manufacturing of active pharmaceutical ingredients. *Pharmaceutical Engineering*, page 34, July/August, 2006.

[145] J. Pekny and G. V. Reklaitis. Towards the convergence of theory and practice: a technology guide for scheduling/planning methodology. *Proceedings of the 3rd Foundations of Computer-Aided Process Operationg*, page 91, 1998.

[146] Intelligen. *"Polymer Resin" SchedulePro example ReadMe file. The full document can be downloaded in PDF format from www.intelligen.com/schedulepro_overview*. 2013.

[147] Intelligen. *SchedulePro functional evaluation version available online at www.intelligen.com/downloads*. 2013.

[148] Intelligen. *"SchedulePro User's Guide," Chapter 4 (Tutorial). The full document can be downloaded in PDF format from www.intelligen.com/demo*. 2013.

Index

Milton Keynes UK
Ingram Content Group UK Ltd.
UKHW031147141024
449569UK00024B/1013